RESEARCH AND PERSPECTIVES IN NEUROSCIENCES

Fondation Ipsen

RESEARCH AND PERSPECTIVES IN NEUROSCIENCES

[illegible] Fondation Ipsen, Paris (France)

[illegible] McGill University, Montreal (Canada).
[illegible] Supérieure, Paris (France).
[illegible] CNRS, UPR 2, Paris (France).
[illegible] INSERM U [illegible] (France)
[illegible] Massachusetts Institute of Technology, Boston (USA).
[illegible] University of [illegible] (Sweden)
[illegible] of Medicine & Dentistry of New Jersey, Piscataway
[illegible]
[illegible] Clinic and Research Foundation, La Jolla (USA)
[illegible] CNRS-INSERM de Pharmacologie-Endocrinologie,
[illegible] (France)
[illegible] des Neurosciences, Paris (France).
[illegible] Collège de France, Institut Pasteur, Paris (France)
[illegible] University of California, Irvine (USA).
[illegible] University of Cambridge, Cambridge (UK)
[illegible] Medical Research Council, Edinburgh (UK).
[illegible] University of California, La Jolla (USA).
[illegible] Collège de France, Paris (France)
[illegible] INSERM U [illegible], Paris (France)
[illegible] URA 372, Marseille (France)
[illegible] INSERM U 259, Bordeaux (France).
[illegible] University of California, [illegible] (USA).
[illegible] McGill University, Montreal (Canada).
[illegible] Washington University Medical School, Saint Louis (USA).
[illegible] INSERM U [illegible], Montpellier (France).
[illegible] University Medical Center, Durham (USA).
[illegible] INSERM U [illegible], Paris (France)
[illegible] Institut Alfred Fessard, CNRS, Gif-sur-Yvette (France).
[illegible] de Neurochimie du CNRS-INSERM U 44 (Strasbourg).

F. Gage Y. Christen (Eds.)

Gene Transfer and Therapy in the Nervous System

Springer-Verlag
Berlin Heidelberg New York
London Paris Tokyo
Hong Kong Barcelona
Budapest

Gage, Fred H., Prof. M. D.
Department of Neurosciences
The University of California
San Diego
La Jolla 92093-0624
USA

Christen, Yves, Ph. D.
Fondation IPSEN pour la Recherche Thérapeutique
30, rue Cambronne
75737 Paris CEDEX 15
France

ISBN-13: 978-3-642-84844-5 e-ISBN-13: 978-3-642-84842-1

DOI: 10.1007/978-3-642-84842-1

Softcover reprint of the hardcover 1st edition 1992

Typesetting: Cicero Lasersatz, 8900 Augsburg

27/3145/5 4 3 2 1 0 – Printed on acid-free paper

Preface

This volume contains the proceedings of the meeting "Gene Transfer and Therapy in the Nervous System" held in Paris on September 23, 1991, and organized by the Fondation Ipsen. This meeting was the second in the series of Colloques médecine et recherche devoted to the neurosciences, the first of which was held on September 24, 1990. The proceedings of this first meeting were published in the same series of books and are entitled Glutamate, cell death and memory (P. Ascher, D. Choi, Y. Christen, eds.). The third Colloque médecine et recherche in neurosciences takes place on November 24, 1992, and will deal with the topic of *Motor and cognitive functions of the prefrontal cortex* (Organizing committee: A. M. Thierry, Paris; J. Glowinski, Paris; P. Goldman-Rakic, New Haven; Y. Christen, Paris). The fourth meeting in this series will be held on October 11, 1993, and is to be entitled "Temporal coding: an area of renewed interest in brain function" (Organizing committee: G. Buzsaki, New York; W. Singer, Frankfurt; R. Llinas, New York; A. Berthoz, Paris; Y. Christen, Paris).

Yves Christen

Introduction

Extraordinary progress has been made over the last few decades in understanding brain function. Advances in molecular biology and the development of molecular genetic tools have provided new insights into normal brain function and human disease in general. New techniques have made it possible to isolate and characterize relevant genes, and initial model systems have been developed to introduce functional wild-type genes into mutant cells to investigate gene function and correct a disease phenotype. The extension of this approach to whole animals – that is, the correction of a disease phenotype in vivo through the use of the functional gene – has come to be called gene transfer and therapy. The validity of gene transfer technology as a tool to increase our understanding of brain function is predicated on the assumption that, by adding to or deleting a gene from the organism or cell, we can begin to understand the function of that gene and its products in the context of the functioning cell and organism. In contrast, gene therapy is based on the assumption that the correction of a disease phenotype can be accomplished either by modification of the expression of a resident mutant gene or by the introduction of new genetic information into defective cells or organism in vivo. Techniques for the ideal versions of gene transfer or therapy, i.e., through site-specific gene sequence correction or replacement, are just beginning to be conceived, but are not yet well developed. A direct method of gene transfer would involve introducing the gene directly into the brain, either as a plasmid or in a vector delivery system such as herpes simplex virus or adenovirus. Some of the progress that has been made is reviewed in two chapters in this volume. The most useful application to date for direct gene delivery has been demonstrated by introducing genes into developing brains. This technique is also summarized in this volume.

Many current models of gene transfer and therapy are really genetic augmentation rather than replacement models; they rely on the development of efficient gene transfer systems to introduce functional, wild-type genetic information into genetically defective cell in vitro and in vivo. To be generally useful to the neurobiologist, these models must combine the availability of efficient delivery vectors for foreign sequences (transgenes) with easy accessibility of suitable target cells or organs and with the development of techniques to introduce the vector stably and safely into those target cells.

There are restrictions on the types and ages of cells that are appropriate for implantation, based on issues of survival and useful functioning for extended periods following implantation. Fetal neuronal tissue has proven to be the most successful donor tissue for brain grafting, and in many cases the success of implants has been dramatic and unequivocal. However, the number of cells and the purity of the specific neuronal phenotype are limiting variables. Further, immunological issues are of critical importance, with immunosuppression sometimes being required for graft survival and function. Finally, the ethical and political debates over the use of fetal tissue hinder clinical application of this experimental tissue approach.

Alternative cell types are being used experimentally. These include glial and peripheral neurons. For example, adrenal chromaffin cells have been used to make and secrete catecholamines. The use of these cells avoids the immunological rejection problem because the host supplies its own cells as an autograft. However, while some moderate success has been documented with fetal adrenal cells, adult chromaffin cells do not survive or function well in the absence of exogenous trophic support. The general strategy of searching the organism for easily accessible cells that can be intracerebrally grafted to substitute for lost brain function requires that:

1. the part of the body donating the cells can do without those cells,
2. the donor cells will survive in the new location, and
3. the donor cells will make and secrete the missing molecule in sufficient quantities to replace the missing or defective function.

This latter requirement may be met by taking advantage of the advances in gene transfer. This approach requires the development of methods to introduce foreign transgenes into appropriate neuronal and other target cells efficiently and functionally in vitro, and to ensure the long-term survival of these cells and continued stable gene expression following intracerebral grafting. The conceptual and methodological development of these general objectives depends on the answers to a variety of specific questions and problems.

The strategy for grafting genetically altered cells to the brain includes the following basic steps:

1. selection of appropriate "transgenes" or genes whose expression is correlated with CNS function or disease;
2. selection and development of suitable and efficient vectors for gene transfer;
3. preparation of donor cells from primary cultures or from established cell lines; and
4. demonstration that the transgene can have a desired phenotypic effect when grafted to the brain of the host animal.

Several examples of the initial attempts to achieve these results are presented in this volume in chapters that deal with a variety of transgenes, cell types and in vivo model systems.

Once a cell, gene, and gene transfer method have been chosen, there are still many issues to resolve in vitro prior to grafting the cells, for example, whether primary cells are susceptible to efficient retroviral infection, whether susceptibility changes with number of passages or doublings, how long primary cells can remain in vitro prior to infection and still survive als well as when they were grafted to the brain, and how long after infection the cell can remain in vitro and still survive well in the brain following grafting. Another important issue for the established cell lines as well as primary cells is whether the confluency of the cells in vitro just prior to grafting is important. It may be important to reach this state of contact inhibition prior to grafting, or, on the other hand, it may be better for the cells to be in a state of growth and division when grafted. Furthermore, under some conditions and with particular cells implanted in certain brain areas, it may be important to determine whether the suspended cells can be immobilized in a matrix which optimizes survival and decreases cell migration. In addition, the question of whether the cells are injected intraparenchymally or intraventricularly, as in a wound cavity, and the age of donor cells and/or host animal may also be critical variables for cell survival and gene expression. Finally, long-term cell survival and stable yet regulatable gene expression will become more important as more experience is gained. Several chapters in this volume address the points raised here and contribute answers to some questions. More importantly, these chapters capture the excitement for the future contributions that gene transfer and therapy will make in our understanding of basic mechanisms of brain function and in developing potential therapies for the brain. Progress will be rapid, and as we gain more knowledge and experience new problems and applications will emerge.

Acknowledgements. The editors wish to thank Mrs. Jacqueline Mervaillie for the organization of the meeting in Paris, Mrs. Mary Lynn Gage for her editorial assistance, Yves Agid, Nicole Le Douarin, Jacques Glowinski, and Robert Williamson for their collaboration as chairmen of the meeting, and Richard Mulligan for his participation.

Fred Gage

Contents

Contributors

Andersen, J. K.
Neuroscience Center, Massachusetts General Hospital East, Charlestown, MA 02129, USA

Anderson, D. J.
Division of Biology 216-76, Howard Hughes Medical Institute, California Institute of Technology, Pasadena, CA 91125, USA

Bartlett, R. J.
Department of Medicine, Division of Neurology, Duke University Medical Center, Durham, NC 27710, USA

Beaudet, L.
Montreal General Hospital Research Institute, Montreal H3G 1A4, Canada

Bebrin, W. R.
Department of Biological Chemistry and Molecular Pharmacology, Harvard Medical School, Boston, MA 02115, USA

Bengzon, J.
Restorative Neurology Unit, Department of Neurology, University Hospital, 22185 Lund, Sweden

Björklund, A.
Department of Medical Cell Research, Biskopsgatan 5, 22362 Lund, Sweden

Breakefield, X. O.
Neuroscience Center, Massachusetts General Hospital East, Charlestown, MA 02129, USA

Brundin, P.
Department of Medical Cell Research, Biskopsgatan 5, 22362 Lund, Sweden

Cadet, J. L.
Department of Neurology, Columbia University, 630 W 168th St., New York, NY 10032, USA

Carlson E.
Department of Neurology, Columbia University, 630 W 168th St.,
New York, NY 10032, USA

Caskey, C. T.
Institute for Molecular Genetics, Baylor College of Medicine,
One Baylor Plaza, Houston, T 77030, USA

Chan, P. H.
Department of Neurology, University of California, San Francisco,
CA 94143, USA

Charron, G.
Montreal General Hospital Research Institute, Montreal H3G 1A4, Canada

Chen, S.
Department of Neurology, University of California, San Francisco,
CA 94143, USA

Chu, L.
Department of Neurology, University of California, San Francisco,
CA 94143, USA

Choi, H. K.
The University of Chicago, Chicago, IL 60637, USA

Clemens, P. R.
Institute for Molecular Genetics, Baylor College of Medicine,
One Baylor Plaza, Houston, TX 77030, USA

Coen, D. M.
Department of Biological Chemistry and Molecular Pharmacology,
Harvard Medical School, Boston, MA 02115, USA

Colin, P.
LGM, CNR, 91198 Gif-sur-Yvette, France

Constantinou, C. D.
Department of Medicine, Division of Neurology,
Duke University Medical Center, Durham, NC 27710, USA

Coté, F.
Montreal General Hospital Research Institute, Montreal H3G 1A4, Canada

Dai, Y.
Molecular Biology and Virology Laboratory, The Salk Institute,
P.O. Box 85800, San Diego, CA 92186-5800, USA

Davar, G.
Neuroscience Center, Massachusetts General Hospital East, Charlestown, MA 02129, USA

DiFiglia, M.
Neuroscience Center, Massachusetts General Hospital East, Charlestown, MA 02129, USA

Dykstra, M. J.
Department of Microbiology, Pathologie and Parasitology, North Carolina, State University, Raleigh, NC 27606, USA

Ebendal, T.
Department of Developmental Biology, Biomedical Center, Uppsala University, 75123 Uppsala, Sweden

Epstein, C. J.
Department of Pediatrics, University of California, San Francisco, CA 94143, USA

Ernfors, P.
Department of Medical Chemistry, Laboratory of Molecular Neurobiology, Karolinska Institute, Stockholm 10401, Sweden

Fahn, S.
Department of Neurology, Columbia University, 630W 168th St, New York, NY 10032, USA

Fink, D. J.
Department of Neurology and VA Medical Center, University of Michigan, Ann Arbor, MI 48105, USA

Friedman, W. J.
Department of Neuroscience and Cell Biology, UMDNJ-Robert Wood Johnson Medical School, 679 Hoes Lane, Piscataway, NJ 08854, USA

Friedmann, Th.
Center for Molecular Genetics and Department of Pediatrics 0634, UCSD School of Medicine, La Jolla, CA 92093, USA

Gage, F. H.
Department of Neurosciences, The University of California, San Diego, La Jolla 92093-0624, USA

Galileo, D. S.
Department of Anatomy and Neurobiology, Washington University School of Medicine, 660 S. Euclied Avenue, St. Louis, MO 63110, USA

Lindvall, O.
Restorative Neurology Unit, Department of Neurology, University Hospital, 22185 Lund, Sweden

Lundberg, C.
Department of Medical Cell Research, Biskopsgatan 5, 22362 Lund, Sweden

Majors, J. E.
Department of Biochemistry and Molecular Biophysics, Washington University School of Medicine, 660 S. Euclid Avenue, St. Louis, MO 63110, USA

Mallet, J.
LGM, CNRS, 91198 Gif-sur-Yvette, France

McKay, R. D. G.
Department of Brain and Cognitive Sciences, MIT, Cambridge, MA 02139, USA

Merlio, J.-P.
Department of Medical Chemistry, Laboratory of Molecular Neurobiology, Karolinska Institute, Stockholm 10401, Sweden

Mizui, T.
Department of Pediatrics, University of California, San Francisco, CA 94143, USA

Naini, A.
Department of Neurology, Columbia University, 630 W 168th St., New York, NY 10032, USA

Naviaux, R. K.
Molecular Biology and Virology Laboratory, The Salk Institute, P.O. Box 85800, San Diego, CA 92186-5800, USA

Olson, L.
Department of Histology and Neurobiology, Karolinska Institute, Stockholm 10401, Sweden

Patterson, P. H.
Biology Division, California Institute of Technology, Pasadena, CA 91125, USA

Persson, H.
Department of Medical Chemistry, Laboratory of Molecular Neurobiology, Karolinska Institute, Stockholm 10401, Sweden

Przedborski, S.
Department of Neurology, Columbia University, 630 W 168th St., New York, NY 10032, USA

Roback, J. D.
The University of Chicago, Chicago, IL 60637, USA

Roses, A. D.
Department of Medicine, Division of Neurology, Duke University Medical Center, Durham, NC 27710, USA

Sanes, J. R.
Department of Anatomy and Neurobiology, Washington University School of Medicine, 660 S. Euclid Avenue, St. Louis, MO 63110, USA

Secore, S. L.
Department of Medicine, Division of Neurology, Duke University Medical Center, Durham, NC 27710, USA

Sharp, N. J. H.
Department of Medicine, Division of Neurology, Duke University Medical Center, Durham, NC 27710, USA

Sternberg, L. R.
Department of Microbiology and Immunology, University of Michigan Medical School, Ann Arbor, MI 48105, USA

Van Camp, S. D.
Department of Food Animal and Equine Medicine, College of Veterinary Medicine, North Carolina State University, Raleigh, NC 27606, USA

Verma, I. M.
Molecular Biology and Virology Laboratory, The Salk Institute, P.O. Box 85800, San Diego, CA 92186-5800, USA

Vos, B.
Neuroscience Center, Massachusetts General Hospital East, Charlestown, MA 02129, USA

Wainer, B. H.
The University of Chicago, Chicago, IL 60637, USA

Won, L.
The University of Chicago, Chicago, IL 60637, USA

Yang, G.
Department of Neurosurgery, University of California, San Francisco, CA 94143, USA

Somatic Cell Gene Therapy

I. M. Verma, R. K. Naviaux, and *Y. Dai*

Summary

Somatic cell gene therapy for the correction of many human genetic diseases is now technically possible. We review several methods and cell types that can be used successfully in gene transfer studies in animals, and discuss their potential promise and limitations in the treatment of human genetic diseases.

Introduction

The advent of modern biology and revolution in medical genetics now makes it feasible to attempt correction of genetic disorders by introducing healthy genes whose product can restore the function. Such a prospect of gene therapy has captured the imagination of both the public and scientists (Miller, 1990a; Verma and Naviaux, 1991; Verma, 1990).

There are essentially two types of gene therapy that can be envisaged at present:
1. germ line gene therapy, and
2. somatic cell gene therapy.

Germ line gene therapy has been successfully carried out in several animals and involves the introduction of a gene into a fertilized egg, so that the progeny is able to make the gene product. Ethical considerations aside, several overwhelming technical problems would preclude consideration of this approach in humans. The major impediment is the inability to control the site of insertion of the healthy gene in the chromosome. A healthy gene could insert in another essential gene, causing unexpected and unwanted side affects. At present, the only viable option is somatic cell gene therapy. The insertion of a single or multiple genes in an individual would be solely to benefit the recipient, without passing it on to future generations. Requirements for somatic cell gene therapy include:
1. correction or circumvention of the defect.
2. no undesirable side effects, and
3. sustained supply of the inserted gene product during the adult life of the individual.

F. H. Gage, Y. Christen (Eds.)
Gene Transfer and Therapy
in the Nervous System

Method of Delivery

The successful outcome of human gene therapy endeavors rests on the ability to efficiently deliver genes into the desired cells. The dream of scientists is to replace the defective gene on the chromosome with the healthy gene in a precise surgical fashion. Because of the recent progress in devising methods of homologous gene replacements, this dream may become reality in the not-too-distant future (Mansour et al. 1988). In the meantime, there are three major modes of inserting genes into cells:

1. chemicals methods, which require the mixing of the DNA with calcium phosphate, DEAE-Dextran or cationic lipids and essentially dumping the mixture on the recipient cells. While this approach is extremely simple, the efficiency of DNA delivery is very poor, in the order of 1 in 10,000 to 100,000 cells. Since the frequency of the stem cells in the bone marrow is at best one in 1,000, the chances of introducing healthy genes in the desired cell type are almost non-existent;
2. physical methods in which DNA can be introduced into cells either by microinjection, which is very tedious, or by electroporation, which can damage the cell. Both of these approaches at present are experimental and unlikely to be useful for introducing DNA in a practical or an efficient manner; and
3. viral vectors. One of the remarkable properties of viruses is their ability to carry genetic information efficiently from cell to cell. Most vectors based on DNA viruses usually have the limitation of the size of the foreign DNA they can carry, because of the constraints of DNA packaging. The other liability is their restricted host range. The advantage of DNA virus vectors is their ability to infect post-mitotic cells. At present, however, retroviral vectors offer the best opportunity to deliver genes into cells, tissues and organs (Miller 1990a; Verma and Naviaux 1991; Verma 1990).

Retrovial Vectors

Retroviruses are a group of RNA-containing viruses which have the ability to cause cancer in experimental animals. Retroviruses have a remarkable and unique lifestyle. They contain RNA as their genetic material which is converted to DNA by the viral enzyme, reverse transcriptase. The viral DNA integrates into the host chromosome as a provirus, which acts as an independent gene, just like insulin, β-globin or other cellular genes. It is then transcribed by host DNA-dependent RNA polymerase, and the viral RNA transcripts are subsequently translated to make viral proteins. The viral RNA associates with the viral proteins to generate new infectious particles. The retroviral DNA can be divided into two components:

1. *cis acting elements:* these include the long terminal repeats (LTR), the packaging sequence (ψ) and the sequences required for the initiation of the second strand of viral DNA. These sequences are essential for the viability of retroviruses; and

2. *trans elements:* the viral coding domain capable of generating *gag, pol* and *env* proteins are required for viral infectivity and can be supplied in *trans*.

One of the concerns with using retroviral vectors is the possibility of generating infectious retroviruses, which have the potential to cause cancer. Once again the unique structure of retroviruses comes to the rescue of scientists (Miller 1990b). After the viral RNA and proteins are synthesized in the cell, they associate with each other to generate an infectious particle. This association requires sequences (ψ sequence) located at the 5' untranslated region of the viral RNA. If these sequences are deleted from the provirus, the synthesis of viral RNA and proteins is unaffected, but the two cannot associate to generate infectious virus particles. Thus, cell lines producing viral replication proteins can be used as factories to generate infectious virus particles containing the desired genes as shown in Figure 3, minimizing the chances of generating infectious viruses. Using packaging cell lines, one can generate recombinant retroviruses containing either an ecotropic envelope (capable of infecting rodent cells) or amphotropic envelope (capable of infecting other species including human). The popular ecotropic cell lines are ψ2, GPE86, CRE, whereas amphotropic cell lines are PA12, PA317, ψAM, AM12 or CRIP (Miller 1990b).

Somatic Tissues

Parameters that would dictate the type of somatic cells for gene insertion would depend on a minimum of the following criterion:
1. the ability to grow and proliferate for extended periods of time, preferably during the entire life-span;
2. easy access to the tissue, and
3. the ability to reintroduce the manipulated cells in the animal. At present the following somatic tissues are being explored as possible candidates which fulfill these requirements.

Bone Marrow

The hematopoietic system is an obvious choice for gene transfer because of the presence of pluripotent stem cells capable of generating all hematopoietic cells of the lymphoid and myeloid lineages, and the extensive experience and well-developed procedures for bone marrow transplantation. Not surprisingly, a major effort of human gene therapy has focussed on the use of hematopoietic cells for sustained gene therapy. The basic procedure involves the following steps:
1. generation of a cell line producing helper free recombinant retroviruses at a titre of a least 5×10^5 virus particles/ml or greater;
2. aspiration of bone marrow (~ 1.0 ml) from a mouse (usually femur or tibia) and cocultivation for 48 hours with irradiated (20 grays) virus-producing cells; and

3. injection of non-adherent bone marrow cells in the tail vein of either lethally irradiated (9.5 grays) or stem cell defective (W/W^v) mice, to allow reconstitution of bone marrow in the recipient mice (Dzierzak et al. 1988; Lim et al. 1989; Belmont et al. 1988; Bodine et al. 1989; Bender et al. 1989; Li et al. 1990). Either the spleen colonies or peripheral blood is analyzed for the presence and expression of recombinant gene. The results to date can be summarized as follows:
1. It is possible to infect pluripotent stem cells with recombinant retroviruses with good efficiency.
2. Secondary recipients can be reconstituted with the bone marrow from donor mice containing foreign genes, indicating that pluripotent stem cells have been infected.
3. The expression of the foreign gene in the various cell types (lymphoid or myeloid) is very inefficient (usually 1–5% of the level of the endogenous gene).

A number of improvements in the infection protocol which utilizes lymphokines like IL3–IL6, etc., have been suggested (Bodine et al. 1989). Furthermore, bone marrow from different species, including primates and man, has been used as a source for introduction of foreign genes via retroviral vectors. The general conclusion is that it is perhaps difficult to infect, with high efficiency, pluripotent cells from higher species, either because they are not cycling or are damaged during the manipulation. Alternatively they may be lacking appropriate viral receptors. Clearly the need is to find improved infection conditions or to enrich pluripotent stem cells.

Skin Cells

There are principally two types of skin cells that can be used:
1. *keratinocytes,* which are primary cells in the protective, outermost layer of the skin that can be cultured *in vitro* and have been used extensively to regenerate skin in burn patients. These cells are, however, very poorly vascularized and may not be ideal for circulating the secreted protein; and
2. *fibroblasts,* which constitute the principle entity of the connective tissue; their task is to provide the gelatinous collagens needed to hold the cells together.

The fibroblasts are buried deep under the epidermis, but are lavishly supplied with blood vessels. They can be cultured *in vitro* and reimplanted under the skin in a collagen matrix. The fibroblasts have been an attractive somatic cell because of the ease of obtaining the tissue and the high efficiency of infection by a wide variety of recombinant retrovirus (Anoon et al. 1987; Palmer et al. 1987; Sorge et al. 1987; Sr. Louis and Verma 1988; Miyanohara et al. 1988; Axelrod et al. 1990). To utilize fibroblasts for gene therapy, the basic protocol involves the isolation of either embryo fibroblasts or by biopsy from adult animals. The cells are then cultured *in vitro* and infected with recombinant retroviruses and, following selection, are embedded in a collagen matrix. The matrix is then implanted in the animal, under the dermis (St. Louis and Verma, 1988; Palmer et al. 1989). To improve the vascularization, a foam gel soaked in fibroblast growth factor can be inserted in the implant. Expression of the secreted protein can be identified in the plasma. Using fibroblast implants, it has been possible to show that human or canine factor IX protein can be secreted in the mouse plasma for 10–15 days (St. Louis and Verma, 1988; Scharfmann et al. 1991). Inability to detected the secreted protein from the implants after a few days has been attributed to:

1. antibodies against the foreign protein,
2. lack of expression from the implant,
3. limited life-span of the fibroblasts in the implant, and
4. rejection of the implant.

Recently we showed that the replacement of CMV promoter by a housekeeping promoter allowed expression of the foreign gene in the implant for at least 60 days (Scharfmann et al. 1991).

Is gene therapy via skin fibroblasts practical? In a mouse containing an implant of half a million cells, we can detect up to 0.2 μg of the factor IX protein/ml of mouse plasma. Considering the mouse has about 3.0 ml of plasma, this translates to about 0.6 μg of factor IX/half a million cells/mouse. The average person has about 5 μg of factor IX protein/ml of plasma and the hemophilia is manifested when the levels of factor IX protein drop lower than 0.5 μg/ml. Since a normal individual has nearly 3 liters of plasma, the minimal total amount of factor IX protein required is 1.5 mg. If half a million cells can make 0.6 μg of factor IX protein, an implant containing one and a half billion (1.5×10^9) cells should be able to secrete sufficient protein to alleviate factor IX deficiency. At present, however, we do not know the longevity of infected fibroblasts in the implants, or the life-span of the implants themselves. The factor IX hemophilia dog model system should prove to be very useful for *in vivo* experimentation.

The skin fibroblasts offer a clear advantage over bone marrow as the target tissue because of the relatively high expression of the inserted gene. Furthermore, if needed, the collagen implant containing infected cells can also be surgically excised. It is worth noting that fibroblasts infected with retroviral vectors capable of expressing and secreting foreign proteins can be implanted at various sites. Recently, fibroblasts containing retroviral vectors which produce neurotransmitter tyrosine hydroxylase were implanted in rat brain. The animals showed altered pharmacological and behavioral characteristics consistent with expression of the gene transferred (Wolff et al. 1989).

Liver Cells

As a large number of inherited metabolic disorders are due to deficiency of enzymes required to function in the liver, it is not surprising that several investigators have been focussing their attention on introducing foreign genes into hepatocytes. Unlike bone marrow or fibroblasts, hepatic tissue is difficult to obtain and hepatocytes cannot be extensively manipulated because they undergo only few cell divisions. Nevertheless, a number of scientists have been able to isolate hepatocytes and infect them with recombinant retroviruses (Ledley et al. 1987; Wolff et al. 1987; Peng et al. 1988; Wilson et al. 1988 a, b; Anderson et al. 1989). Infection efficiencies are rather low (20–25%) and reimplantation of hepatocytes directly into the liver remains a formidable task. A transient reduction in the levels of cholesterol in LDL receptor-deficient Watanabe rabbits has been observe (Wilson et al. 1988 b). Hepatocytes have also been implanted into intraperitoneal or subdermal sites. It has been shown that hepatocytes attached to dextran beads will persist in the peritoneal cavity for up to 60 days (Demetriou et al. 1986). Furthermore, partial correction of the defects in albuminemic rats can be achieved by this procedure. Methods have also been developed whereby direct gene transfer in live can be undertaken. Of particular interest is the method of targeting to the liver cells involving the use of the desired gene complexed with asialoglycoprotein (Wu and Wu 1988). However, only transient expression of the foreign gene has been observed.

Muscle Cells

Muscle constitutes a fairly large portion of the body and is therefore a good choice of tissue for inserting genetic material. The impetus for gene transfer into muscle cells probably arose from two different sets of observations:

1. myoblasts cultured *in vivo* upon injection in the muscles can fuse to form myotubes, and
2. DNA injected directly into muscles was expressed without integrating into the chromosome (Wolff et al. 1990).

Furthermore, expression persisted for many weeks. The protocol for the use of myoblasts as possible somatic tissue for gene delivery would require their isolation and *in vivo* culture. Preliminary results indicate that neonatal or adult mouse myoblasts can be infected by retroviral vectors *in vitro* and that they fuse with endogenous myoblasts to form mytobues. Recently, a number of research groups have shown that C2C12 muscle cells can be infected with recombinant retroviruses producing either human growth hormone or factor IX protein. We have succeeded in obtaining sustained expression of factor IX following transplantation of recombinant retroviral-infected myoblasts in the hind legs of mice. The expression of factor IX is detectable for a period of more than four months.

Endothelial Cells

Due to their direct access to circulating blood, endothelial cells offer an ideal target for delivering genes whose products are secreted. Several groups have shown that endothelial cells isolated from arteries can be infected with retroviral vectors and then seeded in the arteries. Analysis of the arteries showed that persistent expression could be achieved using β-galactosidase as a marker (Zweibel et al. 1989; Wilson et al. 1989; Nabel et al. 1989). The basic protocol involves obtaining endothelial cells from arteries or veins by gentle scraping, and infecting with recombinant retroviruses. The infected cells producing the desired gene product are then seeded in the arteries or veins using balloon catheters. Alternatively, endothelial cells can be directly injected in the arteries. Results to date are promising, but the relative paucity of target cells, and the fact that the endothelial cell lining of the vessel walls is only one cell thick, pose considerable logistical problems. However, local production of biologicals, like tissue plasminogen activator (± PA), can be achieved by even low numbers of endothelial cells.

Perspectives

A question that is frequently posed is what types of genes would be most suitable for gene therapy, since more than 4,500 gene disorders have been identified. From a practical point, it is easy to classify the vast majority of genetic disorders into three groups.

1. *Multifactorial disorders.* These disorders are a consequence of interaction of more than one gene and perhaps other epigenetic factors, for instance, congenital heart lesions, spina bifida, etc. The very rudimentary knowledge of these disorders makes them unlikely candidates for gene therapy.
2. *Chromosomal aberrations.* A number of genetic disorders involve chromosomal abnormalities ranging from microscopic deletions, increased mass and rearrangements. For instance, in Down's Syndrome, the chromosome 21 has three copies instead of the normal two. In the near future, these defects are also unlikely to be candidates for gene therapy.
3. *Single gene disorders.* These genetic disorders caused by a defect in a single gene are the most likely candidates for attempting gene therapy.

We are at the very beginnings of the technology for gene therapy. Clearly we need better vectors which can not only deliver, but express, foreign genes more efficiently. We need more experimentation in animal model systems. We need to constantly monitor and update the safety of the vectors. Despite generating helper-free retroviruses, there is still a slim chance of recombination to generate infectious viruses. There is also the potential of insertion of a recombinant retrovirus in the vicinity of an oncogene which can be inadvertently activated to cause cancer. As in many other medical treatments, the risk and benefit assessment for somatic cell gene therapy for each individual will have to be determined.

The idea of introducing genes into people to correct genetic disorders is nothing less than revolutionary. With the advent of genetic engineering, the practice of medicine is also undergoing a change. It is not uncommon for prospective parents to seek genetic counseling if there is a family history of disease. The technology to screen for prenatal genetic defects from extremely small amounts of tissue (for instance *chorionic villus* in the fetus) with considerable accuracy is rapidly making inroads in the clinics. The success of the human genome initiative will soon provide individuals access to their genetic directory. With such dramatic prospects also comes many fears for the future. There is a very healthy debate on the ethics of human gene therapy. The public has genuine concerns and the right to know. We believe that much of the anticipated fear concerns germ line therapy, where new traits can be introduced and passed on to progeny. The use of somatic tissue, where the inserted gene is solely for the benefit of the recipient, is less likely to arouse public fears. At present, only somatic cell gene therapy is advocated by scientists.

Acknowledgements. A large portion of the manuscript has been adapted from Verma and Naviaux (1991). R. K. Naviaux is supported by a fellowship from the George E. Hewitt Foundation for Medical Research, and Y. Dai is supported by a fellowship from the Muscular Dystrophy Association. This work was also supported by grants from the National Institutes of Health and American Cancer Society to I. M. Verma.

References

Anderson KD, Thompson JA, Dipietro JM, Montgomery KT, Reid LM, Anderson WF (1989) Gene expression in implanted rat hepatocytes following retroviral-mediated gene transfer. Somatic Cell Mol Genet 15:215–227

Anson DS, Hock RA, Austen D, Smith KJ, Brownlee GG, Verma IM, Miller AD (1987) Towards gene therapy for hemophilia B. Mol Biol Med 4:11–20

Axelrod JH, Read M, Brinkhous KM, Verma IM (1990) Phenotypic correction of factor IX deficiency in skin fibroblasts of hemophilic dogs. Proc Natl Acad Sci USA 87:5173–5177

Belmont JW, Macgregor GR, Wager-Smith K, Fletcher FA, Moore KA, Hawkins D, Villalon D, Chang SM, Caskey CT (1988) Expression of human adenosine deaminase in murine hematopoietic cells. Mol Cell Biol 8:5116–5125

Bender MA, Gelinas RE, Miller AD (1989) A majority of mice show long-term expression of a human β-globin gene after retrovirus transfer into hematopoietic stem cells. Mol Cell Biol 9:1426–1434

Bodine DM, Karlsson S, Nienhuis AW (1989) Combination of interleukins 3 and 6 preserves stem cell function in culture and enhances retrovirus-mediated gene transfer into hematopoietic stem cells. Proc Natl Acad Sci USA 86:8897–8901

Demetriou AA, Levenson SM, Novikoff PM, Novikoff AB, Chowdhury NR, Whiting J, Reisner A, Chowdhury JR (1986) Survival, organization, and function of microcarrier-attached hepatocytes transplanted in rats. Proc Natl Acad Sci USA 83:7475–7479

Dzierzak EA, Papayannopoulou T, Mulligan RC (1988) Lineage-specific expression of a human β-globin gene in murine bone marrow transplant recipients reconstituted with retrovirus-transduced stem cells. Nature 331:35–41

Ledley FD, Darlington GJ, Hahn T, Woo SL (1987) Retroviral gene transfer into primary hepatocytes: Implications for genetic therapy of liver-specific functions. Proc Natl Acad Sci USA 84:5335–5339

Li CL, Dwarki VJ, Verma IM (1990) Expression of human α- and mouse/human hybrid β-globin genes in murine hemopoietic stem cells transduced by recombinant retroviruses. Proc Natl Acad Sci USA 87:4349–4353

Lim B, Apperley JF, Orkin SH, Williams DA (1989) Long-term expression of human adenosine deaminase in mice transplanted with retrovirus-infected hematopoietic stem cells. Proc Natl Acad Sci USA 86:8892–8896

Mansour SL, Thomas KR, Capecchi MR (1988) Disruption of the protooncogene int-2 in mouse embryo-derived stem cells: A general strategy for targeting mutations to non-selectable genes. Nature 336:348–352

Miller AD (1990a) Progress toward human gene therapy. Blood 76:271–278

Miller AD (1990b) Retrovirus packaging cells. Human Gene Therapy 1:5–8

Miyanohara A, Sharkey MF, Witztum JL, Steinberg D, Friedmann T (1988) Efficient expression of retroviral vector-transduced human low density lipoprotein (LDL) receptor in LDL receptor-deficient rabbit fibroblasts in vitro. Proc Natl Acad Sci USA 85:6538–6542

Nabel EG, Plautz G, Boyce FM, Stanley JC, Nabel GJ (1989) Recombinant gene expression in vivo within endothelial cells of the arterial wall. Science 244:1342–1344

Palmer TD, Hock RA, Osborne WR, Miller AD (1987) Efficient retrovirus-mediated transfer and expression of a human adenosine deaminase gene in diploid skin fibroblasts from an adenosine deaminase-deficient human. Proc Natl Acad Sci USA 84:1055–1059

Palmer TD, Thompson AR, Miller AD (1989) Production of human factor IX in animals by genetically modified skin fibroblasts: Potential therapy for hemophilia B. Blood 73:438–445

Peng H, Armentano D, Mackenzie-Graham L, Shen RF, Darlington G, Ledley FD, Woo SL (1988) Retroviral-mediated gene transfer and expression of human phenylalanine hydroxylase in primary mouse hepatocytes. Proc Natl Acad Sci USA 85:8146–8150

Scharfmann R, Axelrod JH, Verma IM (1991) Long term *in vivo* expression of retroviral mediated gene transfer in mouse fibroblast implants. Proc Natl Acad Sci USA 88:4626–4630

Sorge J, Kuhl W, West C, Beutler E (1987) Complete correction of the enzymatic defect of type I Gaucher disease fibroblasts by retroviral-mediated gene transfer. Proc Natl Acad Sci USA 84:906–909

St Louis D, Verma IM (1988) An alternative approach to somatic cell gene therapy. Proc Natl Acad Sci USA 85:3150–3154

Verma IM (1990) Gene therapy. Sci Am 262:68–84

Verma IM, Naviaux, RK (1991) Human gene therapy. Curr Opin Gen Dev. 1:54–59

Wilson JM, Jefferson DM, Chowdhury JR, Novikoff PM, Johnston DE, Mulligan RC (1988a) Retrovirus-mediated transduction of adult hepatocytes. Proc Natl Acad Sci USA 85:3014–3018

Wilson JM, Johnston DE, Jefferson DM, Mulligan RC (1988b) Correlation of the genetic defect in hepatocytes from the Watanabe heritable hyperlipidemic rabbit. Proc Natl Acad Sci USA 85:4421–4425

Wilson JM, Birinyi LK, Salomon RN, Libby P, Callow AD, Mulligan RC (1989) Implantation of vascular grafts lined with genetically modified endothelial cells. Science 244:1344–1346

Wolff JA, Yee JK, Skelly HF, Moores JC, Respess JG, Friedmann T, Leffert H (1987) Expression of retrovirally transduced genes in primary cultures of adult rat hepatocytes. Proc Natl Acad Sci USA 84:3344–3348

Wolff JA, Fisher LJ, Xu L, Jinnah HA, Langlais PJ, Iuvone PM, O'Malley KL, Rosenberg MB, Shimohama S, Friedmann T, Gage FH (1989) Grafting fibroblasts genetically modified to produce L-dopa in a rat model of Parkinson's disease. Proc Natl Acad Sci USA 86:9011–9014

Wolff JA, Malone RW, Williams P, Chong W, Acsadi G, Jani A, Felgner PL (1990) Direct gene transfer into mouse muscle in vivo. Science 247:1465–1468

Wu GY, Wu CH (1988) Receptor-mediated gene delivery and expression in vivo. J Biol Chem 263:14621–14624

Zwiebel JA, Freeman SM, Kantoff PW, Cornetta K, Ryan US, Anderson WF (1989) High-level recombinant gene expression in rabbit endothelial cells transduced by retroviral vectors. Science 243:220–222

Delivery of Nerve Growth Factor Within the Rat Brain Using Genetically Modified Cells

M. D. Kawaja and *F. H. Gage*

Summary

Genetic modification of primary skin fibroblasts followed by grafting is a new approach to deliver nerve growth factor within the adult central nervous system. Intracerebral grafts of nerve growth factor-producing fibroblasts have been used to assess the regenerative capacities of select neuronal populations within the damaged adult rat brain. Our data reveal that this trophic factor is a necessary requirement to sustain neurons following axotomy. As importantly, nerve growth factor is needed to induce and promote the regeneration of lesioned axons. Regardless of the cellular or extracellular substrates used by these regrowing axons, the availability of nerve growth factor is a minimum requirement for successful regeneration of adult perturbed axons.

Introduction

A variety of different approaches have been developed to deliver growth factors within the nervous system of adult mammals. These methods of intracerebral delivery have yielded a great wealth of information concerning the effects that different growth factors have upon the survival of neurons following axotomy and the regenerative capacities of lesioned axons. In this review, we will examine the various techniques that have been used to introduce nerve growth factor (NGF), the best-characterized of the neural growth factors, within the nervous system of maturing and adult rodents, as well as describe the results of these investigations. In addition, this review will focus primarily upon the use of cells genetically modified to produce NGF to achieve a focal delivery of this tropic factor within the rodent central nervous system (CNS).

NGF and the mammalian nervous system

There is an intimate association between NGF and several distinct neuronal populations found in both the peripheral and central nervous systems (for reviews, see Thoenen and Barde 1980; Barde 1989; Gage et al. 1991). In particular, peripheral sympathetic and sensory neurons are dependent on target-derived NGF to sustain normal development and maturation (for reviews, see Levi-Montalcini and Angeletti 1968; Thoenen and Barde 1980). These neurons, which express the receptors for NGF, retrogradely transport NGF from their sites of innervation to the cell bodies (Hendry et al. 1974; Paravicini et al. 1975; Stöckel et al. 1975; Johnson et al. 1978; interruption of this target-derived growth factor results in neuronal atrophy and cell death (Purves 1976). The same is true for cholinergic neurons of the basal forebrain. Infusions of iodinated NGF into the hippocampus or cortex result in the labelling of cholinergic neurons of medial septum and basal nucleus, respectively (Schwab et al. 1979; Seiler and Schwab 1984). These same neurons, which also express the receptors for NGF, provide an extensive diffuse innervation to the hippocampus and cortex. Thus, it is apparent that cholinergic neurons residing within the medial septum and nucleus basalis, like peripheral sympathetic and sensory neurons, use NGF receptors in their axons to bind target-derived NGF and transport it in a retrograde manner to the cell bodies. The cholinergic neurons of the basal forebrain, therefore, offer a unique population of CNS neurons to assess the functional aspects of NGF within the adult mammalian brain.

Delivery of NGF: a non-cellular approach

One of the first reports of direct infusion of purified NGF into the rat CNS was conducted by Levi-Montalcini and colleagues. These authors found that, following a one-time injection of NGF into the locus coeruleus of neonatal rats, fibers arising from sympathetic neurons of the superior cervical ganglia grow in an aberrant fashion through the spinal cord and dorsally to the source of exogenous NGF (Menesini-Chen et al. 1978). This observation demonstrated a successful use for infusions of NGF into the CNS to assess either its tropic (the attractiveness of neurites) or trophic (support for developing or damaged neurons) effects. It was not until several years later that researchers began to infuse NGF into the CNS following damage. Between 1986 and 1990, data from several reports revealed varied effects of exogenous NGF within the adult brain, especially upon cholinergic neurons of the basal forebrain. First, using mini-pumps containing reservoirs of purified NGF, investigators showed that intraventricular infusions of NGF near the site of a fimbria-fornix (FF; the pathway through which septal axons reach the hippocampus) lesion can prevent the loss of most cholinergic neurons within the septum of the rat (Hefti 1986; Williams et al. 1986; Kromer 1987; Gage et al. 1988; Montero and Hefti 1988) and non-human primate (Koliatsos et al. 1990; Tuszynski et al. 1990a). Without such exogenous trophic support, axotomized septal cholinergic neurons initially down-regulate the expression of NGF receptor and cholinergic markers, and ultimately a significant proportion (up to 70%) of the population undergoes retrograde degeneration (Hagg et al. 1989; Fischer and Björklund 1991). Recently, Aebischer and his colleagues have taken an alternative approach by using capsules filled with polymers that bind NGF and slowly release it following implantation within the brain (Hoffmann et al. 1990). As shown with minipumps, a significant proportion of neurons can be sustained with the NGF-containing capsules. Second, infusions of NGF have proven useful in reversing age-dependent atrophy of cholinergic perikarya within the nucleus basalis, as well as ameliorating memory deficits (Fischer et al. 1987). Third, NGF infusions into the adult striatum induce neuronal hypertrophy of cholinergic interneurons and increase the expression for mRNA for NGF among cholinergic neurons within the ipsilateral medial septum and nucleus basalis (Gage et al. 1989; Higgins et al. 1989). Finally, intracerebral delivery of exogenous NGF also has a profound effect upon the regenerative capacities of lesioned axons arising from the medial septum. For instance, the degree of axon regeneration and the length of regrowth of septal axons after FF lesions through neural grafts of fetal hippocampus or sciatic nerve increase with infusions of NGF (Hagg et al. 1990; Tuszynski et al. 1990b). Grafting these tissues alone, however, results in fewer and shorter regenerating axons.

As reviewed above, exogenous NGF can be delivered within the brain by means of a miniosmotic pump that releases small amounts of NGF into the ventricular cavity or directly into the parenchyma. One distinct advantage of directly infusing NGF into the brain is that relatively high doses can be delivered and controlled. There are, however, several drawbacks to the long-term use of these miniosmotic pumps. The biological activity of NGF stored within the reservoir may diminish over extended periods. Further, only a finite amount of NGF can be stored within the reservoir, and thus new NGF must be added periodically. The infusion of NGF via a cannula within the ventricle or parenchyma results in chronic damage of the tissue, and may also lead to infection. Finally, mechanical problems may be encountered with miniosmotic pumps. To avert some of these problems, an alternative approach to deliver NGF within the CNS may be through grafting cellular sources of the trophic factor (Gage et al. 1987). The remainder of this review will examine possible sources of NGF-producing cells and their potential use in intracerebral grafting to assess neuron survival and axon regeneration in the damaged brain.

Delivery of NGF: a cellular approach

Possible sources and criteria for their use

There are a few natural cellular sources of NGF that have been or could be used for intracerebral grafting, including mouse sarcoma cells (Cohen et al. 1954; Levi-Montalcini et al. 1954), male mouse submaxillary gland (Levi-Montalcini and Cohen 1960; Caramia et al. 1962), and sciatic nerve (Richardson and Ebendal 1982). Of these tissues, most attention has focused on the use of sciatic nerve segments to serve as bridging material for regrowing CNS axons (David and Aguayo 1981; Benfry and Aguayo 1982; Wendt et al. 1983; Hagg et al. 1990), since this tissue provides both NGF and conducive substrates for growth, i.e., Schwann cells. It is interesting to note that pieces of sciatic nerve have served as grafting material since Tello (1911) observed a sprouting of axons toward these intracerebral implants. Furthermore, tyrosine hydroxylase-immunoreactive axons arising from fetal nigral neurons grafted into the dopaminergically deafferented striatum of adult rats grow into co-grafts of sciatic nerve; such co-transplants, however, do not influence the number of surviving nigral neurons or enhance the outgrowth of tyrosine hydroxylase-positive fibers into the striatum (van Horne et al. 1991). Although these data reveal that intracerebral implants of sciatic nerve can promote tropic and trophic effects upon select neuronal populations, it is important to determine whether other known or unknown growth factors are acting alone or in concert with NGF to elicit neuronal survival and axonal sprouting.

An alternative approach to achieve a focal delivery of NGF within the adult brain is the genetic modification of a homogeneous population of cells in culture followed by implantation within the brain (Gage et al. 1987). Using such genetically modified cells offers two clear advantages over natural sources of NGF: 1) cultured cells that have not been genetically engineered are the perfect control material for biological assessment of the effects of NGF; and 2) effects seen with grafts of genetically modified cells and not with grafts of control cells are most probably attributable to the expression of transgene, NGF. To date, several different cell types have been used to deliver NGF via genetic modification, including immortalized cell lines and primary fibroblasts and astrocytes. These cells must exhibit certain characteristics that make them amenable for gene transfer *in vitro* and additional *in vivo* features following intracerebral implantation. First, accessibility of donor cells is an important feature when choosing a population for genetic modification. For instance, primary adult fibroblasts and keratinocytes are both readily obtainable from skin biopsies. Second, donor cells must also exhibit robust cell division and DNA synthesis in culture; this is to ensure successful incorporation of the transgene by either retroviral infection or transfection (using calcium phosphate or lipofection). Third, the cells must survive within the brain for extended periods of time. To enhance such long-term survival, histocompatibility between the donor cells and the recipient organ is of paramount importance. Fourth, intracerebral grafts of genetically modified cells cannot exhibit continued growth that possibly results in the formation of tumors. Fifth, it would be advantageous if the grafted cells became integrated (structurally or functionally) within the host brain. Finally and most importantly, the cells must continue to synthesize and release the transgene (e.g., NGF) into the surrounding host parenchyma.

Although many types of cells may be used for gene transfer and subsequent intracerebral implantation, data from our laboratory suggest that primary skin fibroblasts are a feasible cellular population for genetic modification and grafting (Kawaja et al. 1991; Kawaja and Gage 1991). Primary fibroblasts are readily obtainable from skin biopsies and can be maintained under standard culture conditions. These cells also grow as rapidly as their immortalized counterparts *in vitro*. Once confluence of primary cells is reached, however, cell division ceases or is greatly retarded. Immortalized cells, on the other hand, continue to grow beyond confluence. As mentioned previously, donor cells must also exhibit gene expression following implantation, as well as stable growth and long-term survival without detrimental effects upon the host brain. A variety of different types of immortalized cells (Rosenberg et al. 1988; Wolf et al. 1988; Ernfors et al. 1989; Cunningham et al. 1991) and primary skin fibroblasts (Kawaja et al. 1992) are able to synthesize and release significant levels of NGF following retroviral-mediated gene transfer *in vitro*. Likewise, these genetically modified cells continue to produce NGF after implantation into the rat brain, as revealed by the percentage of NGF-dependent septal neurons rescued from cell death following axotomy (see below). However, grafts of immortalized cells often exhibit growth patterns that are not conducive to long-term experimentation. For instance, 3T3 cells give rise to intracerebral tumors three to eight weeks after grafting (Horrellou et al. 1990), whereas grafts of 208F cells often atrophy after four weeks, perhaps as a result of histoincompatibility between the host and donor (personal observations). Using primary skin fibroblasts is one means to circumvent problems of tumor formation and immune rejection. Following implantation within the rat brain (in these scenarios the donor and recipient are one and the same animal or the donor and recipient animals are from the same inbred strain), grafts of primary fibroblasts maintain a constant size between three and eight weeks, and the size of these grafts does not appear to change up to six months (Kawaja et al. 1991; Kawaja and Gage 1992). In fact, no spontaneous tumors have yet been observed to arise from implants of primary skin fibroblasts. Morphologically, the fibroblasts appear viable up to six months *in vivo* and possess several typical characteristics of normal skin fibroblasts: spindle shape, elongated nucleus with condensed chromatin and extensive rough endoplasmic reticulum. Finally, fibroblast grafts become structurally incorporated within the host brain via the infiltration of astrocytic processes and capillaries; the extensive vascular plexus found within these grafts resembles host-derived vessels both morphologically (e.g., continuous-type endothelium with a surrounding basal lamina) and functionally (e.g., they posses an intact barrier to blood-borne macromolecules by eight weeks). From these observations, we suggest that genetic modification of homogeneous populations of cells to produce NGF is an alternative approach for delivering NGF within the adult CNS. Furthermore, of the many types of cells that can undergo successful gene transfer and subsequent intracerebral grafting, primary skin fibroblasts offer the best characteristics, both *in vitro* and *in vivo*.

Intracerebral grafts of NGF-producing fibroblasts: Neurotrophic/tropic effects

Over the past few years, immortalized and primary fibroblasts genetically modified to produce NGF have been used to address questions concerning neuronal survival and axon regeneration following perturbation. Using the model of lesion-induced neuronal degeneration within the septo-hippocampal system, we grafted genetically modified primary fibroblasts to determine whether these cells would be effective in their production and secretion of NGF *in vivo* (Kawaja et al. 1992). Two different approaches have been taken to graft NGF-producing primary fibroblasts into rat brains following unilateral FF ablations. First, cells were implanted directly within the septal parenchyma ipsilateral to the lesion. Second, fibroblasts were suspended in rat tail collagen *in vitro* and pieces of this matrix material were subsequently implanted within the lesion cavity. Four weeks after unilateral FF ablation and septal implantation of primary fibroblasts, a significant proportion (~ 75%) of NGF receptor-positive septal neurons were sustained with NGF-producing fibroblasts, whereas only 45% of septal immunoreactive neurons survived axotomy with grafts of non-infected control cells. When grafts of collagen and NGF-producing fibroblasts were implanted within the lesion cavity, a significant percentage of septal cholinergic/NGF receptor-positive neurons was sustained at four and eight weeks after axotomy (~ 65%). In comparison to the proportion of neurons sustained with NGF-producing rat 208F and mouse NIH-3T3 cells (i.e., > 90%, Rosenberg et al. 1988; Strömberg et al. 1990), that achieved with primary skin fibroblasts is smaller but significant (65–75%). Since the rates of NGF release appear to be comparable for both immortalized 208F cells and primary skin fibroblasts, differences in cell growth following intracerebral grafting may be one reason for the disparity. Intracerebral grafts of primary skin fibroblasts do not exhibit continued growth (Kawaja et al. 1991), whereas immortalized cells multiply rapidly and often give rise to tumors within a few weeks after implantation (Horellou et al. 1990). Therefore, the greater percentage of axotomized cholinergic cells sustained with grafts of immortalized fibroblasts versus primary fibroblasts may be due directly to a higher rate of cell growth, which in turn results in higher levels of collective NGF production. From these data, we propose that the proportion of lesioned septal neurons saved is dependent upon the available dose of NGF, such that high levels of NGF released from either miniosmotic pumps or grafts of immortalized cells sustain a larger number of axotomized cholinergic cells than do lower levels of NGF likely to be released from grafts of primary skin fibroblasts.

Not only do grafts of NGF-producing primary fibroblasts prevent a significant proportion of septal neurons from retrograde degeneration, but they also induce axon sprouting. To more closely assess the tropic effects of such grafts upon cholinergic neurons of the rat basal forebrain, we grafted suspensions of NGF-producing primary fibroblasts within the adult striatum (Kawaja and Gage, 1991). Between one and three weeks following implantation, NGF receptor-positive axons arising from the adjacent nucleus basalis grew towards and surrounded grafts of NGF-producing fibroblasts. Grafts of control non-infected cells, however, did not elicit the same response. By eight weeks, the NGF-producing grafts were filled with NGF receptor-immunoreactive profiles; the control grafts lacked this pattern of immunostaining. At the electron microscope level, numerous unmyelinated axons were enveloped within glial processes, especially those of reactive astrocytes, in grafts of NGF-producing fibroblasts only. Furthermore, these axons were exclusively associated with glial elements and not found within the extracellular matrix of the grafts. Although control grafts possessed reactive astrocytic processes, axons were not observed anywhere within these grafts. From these data, we conclude that

1. reactive astrocytes can act as conducive substrates for growing axons; and
2. only in the presence of elevated levels of NGF will permissive substrates (e.g., astrocytes) support axon growth by NGF-sensitive neurons.

Our observations that intracerebral grafts of NGF-producing skin fibroblasts can induce axonal sprouting lead us then to ask whether these grafts could also promote axon regeneration among NGF-sensitive neuronal populations following axotomy. And if regenerating axons could grow through NGF-producing grafts, would they be able to reinnervate target sites within the adult brain? To address these questions, we again focused our attention on the septo-hippocampal model (Kawaja et al. 1992). Following unilateral FF ablations, collagen-fibroblast matrices were grafted within the lesion cavity to assess the importance of NGF on the regenerative capacities of rat septal neurons after axotomy. First, we found that NGF can promote the regeneration of septal axons, since NGF-producing grafts possessed large numbers of acetylcholinesterase (AChE)-positive axons as early as four weeks after grafting. Grafts of control cells, however, usually lacked cholinergic axons. At the electron microscope level, unmyelinated axons observed within NGF-producing grafts use many different substrates for growth, including astrocytes and components of the extracellular matrix, including the basal lamina of capillaries and collagen. Grafts of control fibroblasts possessed the same cellular and matrix substrates but contained only a very small population of axons, probably of peripheral sympathetic origin. Finally, AChE-positive axons growing through NGF-producing grafts provided a topographically organized input to the deafferented hippocampal dentate gyrus. Further, regenerating septal axons terminated predominantly on the dendritic processes of granular neurons of the dentate gyrus. On the other hand, the dentate gyrus ipsilateral to grafts of control non-infected fibroblasts remained devoid of AChE-positive fibers. From these results, we conclude that the availability of NGF is a necessary requirement to promote axon regeneration and cholinergic reinnervation of granular neurons by lesioned septal neurons. The presence of many permissive substrates (e.g., astrocytes, basal lamina and collagen) alone, however, is not sufficient to induce axon regrowth from adult septal neurons.

Conclusions

Primary skin fibroblasts can be used successfully for gene transfer and subsequent intracerebral implantation to achieve an *in vivo* delivery of NGF. Following grafting within the lesioned rat brain, the production of NGF by these cells can sustain a significant proportion of axotomized septal neurons, which without such trophic support would undergo retrograde degeneration. Also, NGF production induces the regrowth of damaged axons arising from cholinergic basal forebrain neurons. These regenerating axons can use a number of different cellular and extracellular matrix components as substrates for growth, but only in the presence of NGF produced by genetically modified fibroblasts will axon regrowth be induced and promoted. Although these regenerating axons are also able to reinnervate granular neurons, it remains to be shown what role NGF plays in the guidance of axons to appropriate target sites and the reformation of synaptic contacts.

References

Barde Y-A (1989) Trophic factors and neuronal survival. Neuron 2:1525–1534

Benfry M, Aguayo A (1982) Extensive elongation of axons from rat brain into peripheral nerve grafts. Nature 296:150–152

Caramia F, Angeletti PU, Levi-Montalcini R (1962) Experimental analysis of the mouse submaxillary salivary gland in relationship to its nerve growth factor content. Endocrinology 70:915–922

Cohen R, Levi-Montalcini R, Hamburger V (1954) A nerve growth-stimulating factor isolated from sarcomas 37 and 180. Proc Natl Acad Sci USA 40:1014–1018

Cunningham LA, Short MP, Vielkind U, Breakefield XO, Bohn MC (1991) Survival and differentiation within the adult mouse striatum of grafted rat pheochromocytoma (PC12) genetically modified to express recombinant β-NGF. Exp Neurol 112:174–182

David S, Aguayo A (1981) Axonal elongation into peripheral nervous system "bridges" after central nervous system injury in adult rats. Science 214:931–933

Ernfors P, Ebendal T, Olson L, Mouton P, Strömberg I, Persson A (1989) A cell line producing recombinant nerve growth factor evokes growth responses in intrinsic and grafted central cholinergic neurons. Proc Natl Acad Sci USA 86:4756–4760

Fischer W, Björklund A (1991) Loss of AChE-- and NGFr-labelling precedes neuronal death of axotomized septal-diagonal band neurons: Reversal by intraventricular NGF infusion. Exp Neurol 113:93–108

Fischer W, Wictorin K, Björklund A, Williams LR, Varon S, Gage FH (1987) Amelioration of cholinergic neuron atrophy and spatial memory impairment in aged rats by nerve growth factor. Nature 329:65–68

Gage FH, Wolff JA, Rosenberg MB, Xu L, Yee JL, Shults C, Friedmann T (1987) Grafting genetically modified cells to the brain: possibilities for the future. Neuroscience 23:795–807

Gage FH, Batchelor P, Chen KS, Chin D, Higgins GA, Koh S, Deputy S, Rosenberg MB, Fischer W, Björklund A (1989) NGF-receptor re-expression and NGF-mediated cholinergic neuronal hypertrophy in the damaged adult neostriatum. Neuron 2:1177–1184

Gage FH, Tuszynski MH, Chen KS, Fagan A, Higgins GA (1991) Nerve growth factor in the central nervous system. In: Bothwell M (ed) Trophic factors in the central nervous system. Springer-Verlag Berlin, pp 71–93

Hagg T, Fass-Holmes B, Vahling HL, Manthorpe M, Conner JM, Varon S (1989) Nerve growth factor (NGF) reverses axotomy-induced decreases in choline acetyltransferase, NGF receptor and size of medial septum cholinergic neurons. Brain Res 505:29–38

Hagg T, Vahlsing HL, Manthorpe M, Varon S (1990) Septohippocampal cholinergic axonal regeneration through peripheral nerve bridges: Quantification and temporal development. Exp. Neurol 109:153–163

Hefti F (1986) Nerve growth factor promotes survival of septal cholinergic neurons after fimbrial transections. J Neurosci 8:21155–2162

Hendry IA, Stöckel K, Thoenen H, Iversen LL (1974) The retrograde axonal transport for nerve growth factor. Brain Res 68:103–121

Higgins GA, Koh S, Chen KG, Gage FJ (1989) NGF induction of NGF receptor gene expression and cholinergic neuronal hypertrophy within the basal forebrain of the adult rat. Neuron 3:247–256

Hoffman D, Wahlberg L, Aebischer P (1990) NGF released from a polymer matrix prevents loss of ChAT expression in basal forebrain neurons following a fimbria-fornix lesion. Exp Neurol 110:39–44

Horellou P, Brundin P, Kalén P, Mallet J, Björklund A (1990) In vivo release of DOPA and dopamine from genetically engineered cells grafted to the denervated rat striatum. Neuron 5:393–402

Johnson EM, Andres RY, Bradshaw RA (1978) Characterization of the retrograde transport of nerve growth factor (NGF) using high specific activity [^{125}]NGF. Brain Res 150:319–331

Kawaja MD, Gage FH (1992) Morphological and neurochemical features of cultured primary skin fibroblasts of Fischer 344 rats following striatal implantation. J Comp Neurol 317:102–116

Kawaja MD, Gage FH (1991) Reactive astrocytes are substrates for the growth of adult CNS axons in the presence of elevated levels of nerve growth factor. Neuron 7:1019–1030

Kawaja MD, Fagan AM, Firestein BL, Gage FH (1991) Intracerebral grafting of cultured autologous skin fibroblasts into the rat striatum: An assessment of graft size and ultrastructure. J Comp Neurol 307:695–706

Kawaja MD, Rosenberg MB, Yoshida K, Gage FH (1992) Somatic gene transfer of nerve growth factor promotes the survival of axotomized septal neurons and the regeneration of their axons in adult rats. J Neurosci 12:2849–2864

Koliatsos VE, Nauta HJW, Clatterbuck RE, Holtzmann DM, Mobley WC, Price DL (1990) Mouse nerve growth factor prevents degeneration of axotomized basal forebrain cholinergic neurons in the monkey. J Neurosci 10:3801–3813

Kromer LF (1987) Nerve growth factor treatment after brain injury prevents neuronal dealth. Science 235:214–216

Levi-Montalcini R, Angeletti PU (1968) Nerve growth factor. Physiol Rev 8:534–569

Levi-Montalcini R, Cohen S (1960) Effects of the extract of the mouse submaxillary salivary glands on the sympathetic system of mammals. Ann NY Acad Sci 85:324–341

Levi-Montalcini R, Meyer H, Hamburger V (1954) *In vitro* experiments on the effects of mouse sarcomas 180 and 37 on the spinal and sympathetic ganglia of the chick embryo. Cancer Res 14:49–57

Menesini-Chen MG, Chen JS, Levi-Montalcini R (1978) Sympathetic nerve fiber ingrowth in the central nervous system of neonatal rodent upon intracerebral NGF injections. Arch Ital Biol 116:53–84

Montero CN, Hefti F (1988) Rescue of lesioned septal cholinergic neurons by nerve growth factor: Specificity and requirement for chronic treatment. J Neurosci 8:2986–2999

Paravicini U, Stoeckel K, Thoenen H (1975) Biological importance of retrograde axonal transport of nerve growth factor in adrenal neurons. Brain Res 84:279–291

Purves D (1976) Functional and structural changes in mammalian sympathetic neurons following colchicine application to post-ganglionic nerves. J Physiol (Lond) 259:159–175

Richardson PM, Ebendal T (1982) Nerve growth activities in rat peripheral nerve. Brain Res 246:57–64

Rosenberg MB, Friedmann T, Robertson RC, Tuszynski M, Wolff JA, Breakefield XO, Gage FH (1988) Grafting genetically modified cells to the damaged brain: restorative effects of NGF expression. Science 242:1575–1578

Schwab ME, Otten U, Thoenen H (1979) Nerve growth factor (NGF) in the rat CNS: Absence of specific retrograde axonal transport and tyrosine hydroxylase induction in locus coeruleus and substantia nigra. Brain Res 168:473–483

Seiler M, Schwab ME (1984) Specific retrograde transport of nerve growth factor (NGF) from neocortex to nucleus basalis in the rat. Brain Res 300:33–3

Stöckel K, Schwab ME, Thoenen H (1975) Specificity of retrograde transport of nerve growth factor (NGF) in sensory neurons: A biochemical and morphological study. Brain Res 89:1–14

Strömberg I, Wetmore CJ, Ebendal T, Ernfors P, Persson H, Olson L (1990) Rescue of basal forebrain cholinergic neurons after implantation of genetically modified cells producing recombinant NGF. J Neursoci Res 25:405–411

Tello F (1911) La influencia del neurotropismo en la regeneracion de las centros nerviosos. Trab Lab Biol Univ Madr 9:123–159

Thoenen H, Barde Y-A (1980) Physiology of nerve growth factor. Physiol Rev 60:1284–1334

Tuszynski MH, Buzsaki G, Gage FH (1990a) Nerve growth factor infusions combined with fetal hippocampal grafts enhance reconstruction of the lesioned septohippocampal formation. Neursocience 36:33–44

Tuszynski MH, U HS, Amaral DG, Gage FH (1990b) Nerve growth factor infusion in primate brain reduces lesion-induced cholinergic neuronal degeneration. J Neursoci 10:3604–3614

van Horne CG, Strömberg I, Young D, Olson L, Hoffer B (1991) Functional enhancement of intrastriatal dopamine-containing grafts by the co-transplantation of sciatic nerve tissue in 6-hydroxydopamine-lesioned rats. Exp Neurol 113:143–154

Williams LR, Varon S, Peterson GM, Wictorin K, Fischer W, Björklund A, Gage FH (1986) Continuous infusion of nerve growth factor prevents basal forebrain neuronal death after fimbria-fornix transection. Proc Natl Acad Sci USA 83:9231–9235

Wolf D, Richter-Landsberg C, Short MP, Cepko C, Breakefield XO (1988) Retrovirus-mediated gene transfer of β-nerve growth factor into mouse pituitary line AtT-20. Mol Biol Med 5:43–59

A Potential Approach to Gene Therapy of Cancer

Th. Friedmann

Oncogenes and Tumor Suppressor Genes

Enormous progress has recently been made in the characterization of the biochemical and molecular events associated with the control of the mammalian cell cycle and with cell replication. Some of the normal cellular genes engaged in such regulatory functions have been shown to lead to, or be associated with, aberrant growth of cells and to the development of the neoplastic changes in the cells. There are at least two kinds of genes associated with the aberrant growth properties of tumor cells (Bishop 1991). Mutated, activated forms of the normal cellular proto-oncogenes are able to cause neoplastic transformation of cells by disrupting the normal controlled mechanisms of cell replication and growth. Their gene products can be imagined to provide positive stimuli to cellular growth. The presence of a single mutated allele is sufficient to cause cell transformation, and the continued presence and expression of the other, wild-type allele is insufficient for the maintenance of the normal cellular phenotype (Fig. 1). Potential therapeutic approaches toward the genetic reversal of these kinds of defects might include interference with translation of oncogene messenger RNAs by the introduction of anti-sense sequences or ribozymes specific for the oncogene transcript, the introduction of transdominant wild-type alleles or the introduction of non-allelic genes encoding functions that interfere with the transforming function of the oncogene (Fig. 2).

Other normal cellular genes are thought to regulate the cell replication machinery by entirely different mechanisms. The existence of this class of genes has been inferred for several decades from studies by Harris (1988) and others, in which hybrid cells were formed by the fusion of human tumor cells with normal human cells (Harris, 1988). The studies demonstrated that, almost invariably, the neoplastic properties expressed in the tumor cell were suppressed in the hydrid cell (Fig. 3). A number of possible mechanisms could be responsible for this effect, including hybridization-induced shifts in the state of differentiation of the tumor cells or the action of a growth suppressing function in the normal cell provided to the hybrid by the normal. Further evidence for the existence of some kinds of growth suppressor functions in human cancers came from the growing evidence being accumulated by a number of cancer cytogeneticists that human cancer cells often demonstrate consistent and reproducible losses of genetic material from specific chromosomes. Furthermore, in

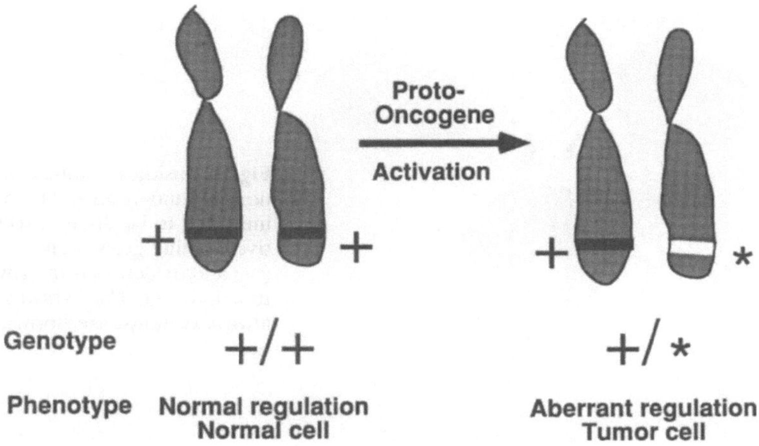

Fig. 1. Action of a dominantly acting transforming oncogene. The presence of two wild type alleles of a normal cellular protooncogene (+) permit normal cellular growth and replication properties. When one of the alleles is mutated to produce a transforming form of the gene, i.e., an oncogene (*), regulation of the growth and replication properties of cell is disrupted and the cell assumes the tumorigenic phenotype.

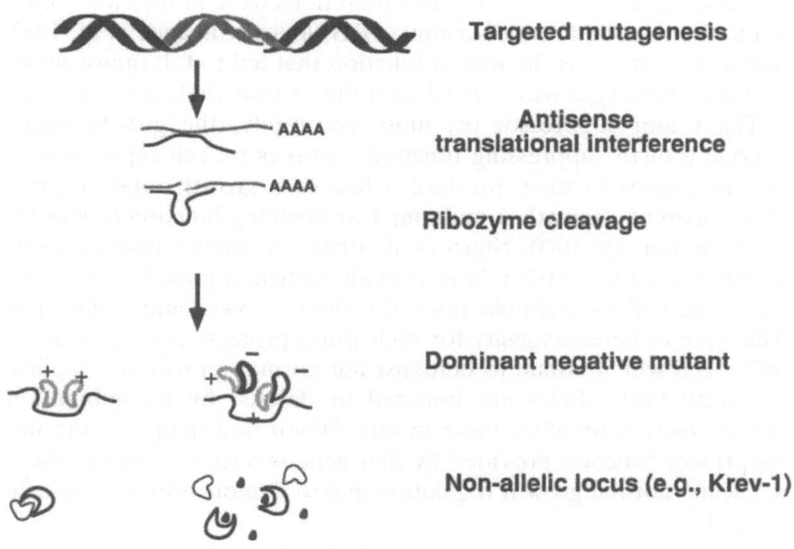

Fig. 2. Potential approaches to inactivation of transforming oncogenes. The sequence of a mutated protooncogene (stippled region of the double helix) might be corrected by targeted mutagenesis methods. Alternatively, translation of the mRNA for the oncogene might be suppressed by antisense sequences or the messenger RNA itself might be cleaved and thereby inactivated by the action of a sequence-specific ribozyme. The action of the oncogene product might be abrogated by the presence of gene products produced by other mutant forms of the same oncogene that prevent multimerization of the transforming protein or its interaction with target molecules. Finally, the action of the transforming oncogene product might be prevented by the presence of proteins produced by non-allelic genes that interfere with cellular transformation.

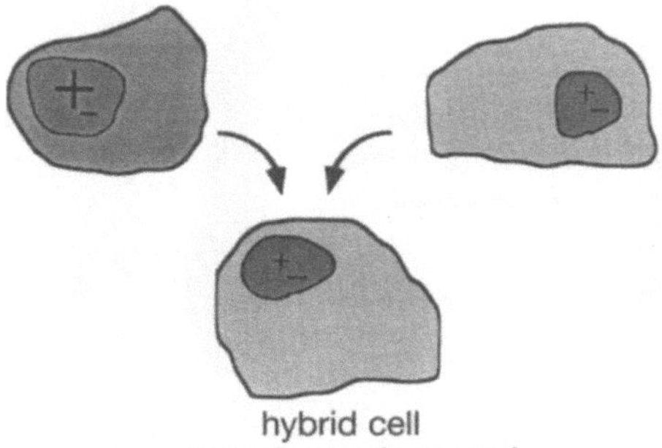

Fig. 3. Fusion of human tumor cells with normal human cells. The tumor cell can be imagined to be driven excessively by positively acting growth and replication stimuli (+) and/or deficient in growth suppressing functions (–). The hybrid cells almost always demonstrate normal growth properties.

an analysis of the epidemiological features of the disease retinoblastoma, Knudson (1971) first suggested a model of cancer pathogenesis in which two genetic defects, one carried in the germ line and the other representing a somatic cell mutation, were required for the development of disease. When it was later determined that the two mutations were in the two alleles of the same gene, the so-called retinoblastoma (RB) locus (Cavanee et al. 1983), it became apparent that it was the loss of function that led to full tumor development and that the phenotype was normal until the second allele became mutated or lost.

The tumor suppressor or, more accurately, the growth suppressor genes encode growth suppressing function – brakes on cell replication and growth – and mutations in them produce a loss of a growth retarding function rather than a gain of a growth stimulating transforming function (Klein 1987; Marshall 1991; Stanbridge 1990; Nigro et al. 1989). A single wild-type allele of a tumor suppressor gene is still able to provide sufficient growth regulatory function to allow the cell to maintain normal cellular growth and replication properties. The state of heterozygosity for such genes protects cells from aberrant growth, and a cell will continue to demonstrate normal growth and replication properties until both alleles are mutated or deleted by recombination, mutation, nondisjunction or other mechanisms. When that happens, the normal growth suppressor function provided by that gene is missing, causing the affected cells to escape normal growth regulation and to demonstrate a neoplastic phenotype (Fig. 4).

A genetic approach to the correction of this kind of defect might thereoretically come through the restoration of the state of heterozygosity for the gene (Fig. 5). If one were to restore the function of one single wild-type allele – i.e., restore the state of heterozygosity – of this kind of gene in cells that had become neoplastic due to loss to heterozygosity of the gene, might it be possible to restore at least some of the normal growth properties of the cell by restoring a state of heterozygosity for that gene? The likelihood would, at first glance, seem to be minimized by the current realization that most neoplastic changes in human and other mammalian cells are likely to be caused by the

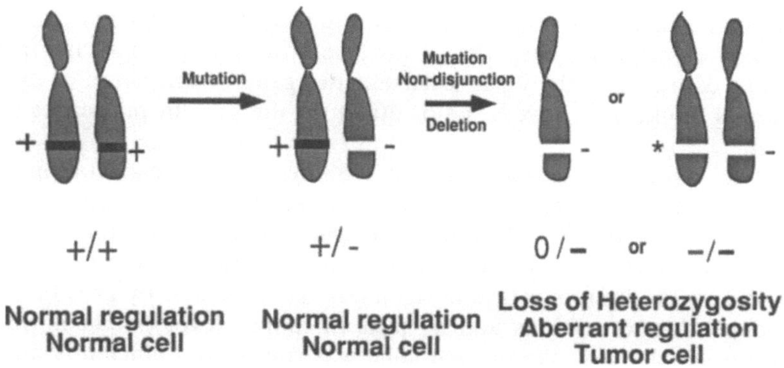

Fig. 4. Mechanism of action of tumor suppressor genes. The presence of both normal alleles of a tumor suppressor gene (+) hold the cell in a state of normal growth and replication. When only one allele is mutant (–), the remaining wild-type allele provides sufficient function to maintain normal cellular growth properties to permit continued normal growth properties even in the presence of the mutated allele. However, when the second allele is also lost or becomes mutated, the cell is totally deprived of the growth suppressing functions of the gene, and the cell assumes a tumorigenic phenotype.

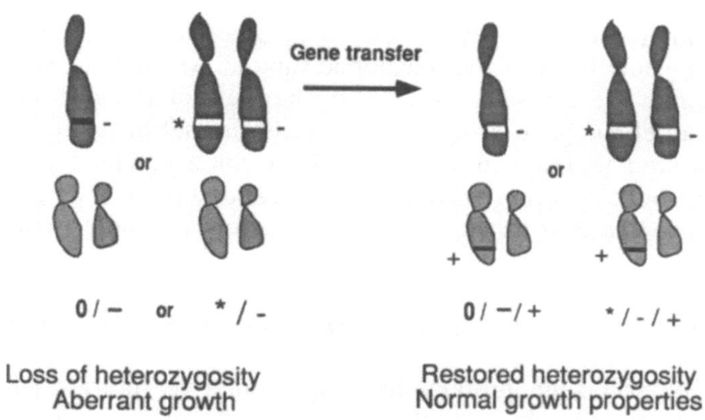

Fig. 5. Restoration of normal growth properties of a tumor cell through restoration of tumor suppressor gene function. A normal allele of the gene is introduced, usually into an ectopic chromosomal site, resulting in a functionally restored state of heterozygosity. The phenotype of the cell reverts from the tumor phenotype to the normal phenotye.

accumulation, possibly in a sequential manner, of a number of mutations in both the oncogenes and the tumor suppressor genes, whose products then cooperate and interact to deregulate the growth properties of a cell and convert it into a cell with neoplastic properties. The most complete clarification of this phenomenon has been provided by the recent work of Vogelstein and his colleagues on the mutations responsible for human colon carcinoma (Fearon and Vogelstein 1990). Cells of the intestinal epithelium are transformed from normal to malignant by a cascade of interacting mutations in oncogenes and in tumor suppressor genes. This requirement for multiple defects for the development of tumorigenicity suggests that the genetic correction at any one single step is unlikely to have a major effect on the growth properties of the tumor cell.

Our laboratory demonstrated in 1988 that such an expectation might not be fully justified (Haung et al. 1988). We found that the restoration of function of a single wild-type RB allele into osteosarcoma cells, i.e., the restoration of the state of heterozygosity, led to a slowdown of the growth properties of the cells and a markedly reduced tumorigenicity, as measured by growth in anchorage-independent conditions and by tests of tumor development in nude mice. In our original study, the normal gene was transferred efficiently and stably into the tumor cells by the use of a retroviral vector encoding the normal human RB allele. Similar results have been obtained with prostatic carcinoma cells also deficient in the RB function and with osteosarcoma cells deficient in the p53 tumor suppressor gene (Bookstein et al. 1990; Chen et al. 1990). However, more recent studies with the RB system have indicated that tumors may continue to develop even after the cells are infected with RB-expressing retroviruses, especially when the cells are introduced into the eye rather than into subcutaneous sites. However, in these recent studies, it is not yet completely clear that the wild-type RB gene product continues to be expressed in the tumor cells or whether deleted or otherwise mutated forms of RB had been selected in the newly developed tumors. Much work is still required to characterize these results fully. A further test of the restoration of heterozygosity model has been provided by Vogelstein and his colleagues, who have demonstrated that the in vitro growth of colon carcinoma cells can be slowed by introducing genes encoding a normal copy of DCC, a gene that constitutes one of the earliest targets in the development of colon carcinoma.

p53 and tumor suppression – human T cell leukemia

The p53 gene product has been found to interact with a number of viral transforming genes such as the SV40 T antigen, adenovirus E1B and the E6 gene product of human papilloma virus (HPV) 16 and 18 (Levine 1990). The earliest evidence about the mechanisms of action of p53 suggested that it was in fact a transforming oncogene, since the p53 genes initially isolated were able to cooperate with activated *ras* oncogene to transform rat embryo cells. However, it soon became clear that the genes initially isolated were mutant genes and that the wild-type gene was unable to cooperate with activated *ras*. The gene therefore was acting as a tumor suppressor gene (Lane and Benchimol 1990). This expanded understanding of the biology of p53 function led to the suggestion that these viral oncoproteins transform cells at least partly by interfering with the normal growth regulatory functions of cellular growth suppressor functions, of which p53 seems to be one.

The p53 tumor suppressor is the gene most commonly found to be mutated in human cancer (Nigro et al. 1989). The clearest role of p53 in human cancer in the familial cancer syndrome described by Li and Fraumeni (Malkin et al. 1990). In this disease, there are underlying germ-line p53 mutations to which are added somatic mutations, leading to a familial cancer syndrome involving cancers of the muscle, breast, kidney and other organs. p53 is also often mutated in many sporadic human cancers, including human hepatocellular carcinoma (Hsu et al. 1991), in some cases of T cell leukemia (Cheng and Haas 1990) and other hematopoietic malignancies, in breast adenocarcinoma (Bartek et al. 1990) and in glioblastomas (Mercer et al. 1990). One implication of this finding is that p53 carries out vital cellular functions common to many cells, and that mutations in this gene and possibly in other central tumor suppressor genes are determinant factors in the conversion of a normal cell into a transformed, tumorigenic cell.

We have been studying the effect of the expression of exogenous wild-type p53 or the phenotype of a number of tumor cells and have determined that, under some conditions, restoration of expression of p53 in cells of human glioblastomas, hepatocellular carcinomas, breast cancers and T cell leukemias can produce changes in cellular growth properties and tumorigenicity suggestive of a suppression of the cancer phenotype (Friedmann et al., unpublished results). Our approach has been to infect tumor cells known to express either no p53 or mutant p53 alleles with a retroviral vector expressing the wild-type gene and to determine the effects of restored p53 expression on morphology, growth rate, cell cycle events and tumorigenicity in anchorage independent cultures and in nude mice. Infection of human Be-13 T leukemic cells with a retroviral vector expressing the wild-type p53 gene causes cells to reduce their growth rate and to lose their ability to grow in methyl cellulose significantly. Furthermore, inoculation of the genetically corrected cells into athymic mice results in a marked reduction in the appearance of tumors (Chang et al. 1992) Table 1 summarizes our results on the effects of restored p53 expression on the tumorigenicity of neoplastic human T cells (Be-13) both in vitro and in vivo.

These results are reminiscent of phenotypic changes produced in tumor cells by the expression of virally transduced RB transgene in osteosarcoma cells or

Table 1. Tumorigenic properties of p53-deficient Be-13 cells before and after retroviral vector transduction. Be-13 cells were infected with either retroviral vector expressing the firefly luciferase (luc) or the wild-type human p53 gene. Tumors were evaluated 28 days after the subcutaneous injection of 10^7 luc-expressing cells in the flank and p53-expressing cells in the right flank of the same mice. Colonies in methyl cellulose were evaluated after two weeks of growth in culture

Cells	In vitro colony formation in methyl cellulose (colonies/10^3 cells)	Tumors in athymic mice (number of animals with tumors/number animals injected)
Be-13	ND	2/2
Be-13-luc	489 + 48	6/6
Be-13-p53 #1	60 + 9	0/7
Be-13-p53 #2	78 + 12	0/2

prostatic carcinoma cells or the p53 transgene in osteosarcoma cells, as described above. The most simplistic interpretation of these and additional recent results is that the restoration of the state of heterozygosity of p53 in these tumor cells suppresses aspects of the tumor phenotype as a result of p53 expression. However, because of the still major uncertainties regarding the mechanisms of action of p53 and its role in the control of growth, replication and cell cycle events, such a facile interpretation of the effects of p53 transgene expression in tumor cells may not be entirely correct or complete. Rather than simply reversing a genetic deficiency of the p53 gene product by returning the target cell to a normally regulated state of heterozygosity, the apparently tumor suppressor effects of restored but unregulated p53 expression may be non-specific and mediated by alterations in the state of cellular differentiation, by complicated and fortuitous changes in the relative amounts of growth-stimulatory and growth retarding functions or even by toxic effects of overexpression of growth suppressor gene products in the recipient cells. Nevertheless, even though transduced T cell leukemia cells suffer no overt or serious damage after infection with the vector, they do demonstrate major changes in some aspects of the tumor phenotype, i.e., their growth rate and their ability to produce tumors in ummuncompromised mice.

The ability to suppress the tumorigenic properties of cells by infecting them in vitro with tumor suppressor-expressing viral vectors is an important proof of principle, but it is only the first in an arduous series of steps required to transform such findings into therapy of existing tumors in vivo. There are many technical hurdles along that path that must be overcome, including the preparation of exceedingly large amounts of virus at much higher titers than are generally feasible at the present time and, more importantly, the development of methods to target the vector to the appropriate tissue or organ in vivo. Direct introduction of vector into tumors or into the vascular supply of tumors is certainly possible, but ideally one would like to introduce biological targetting to the gene transfer techniques so as to minimize unwanted genetic modification of normal cells. While a number of biological targetting approaches have been envisioned, such as introduction of elements into the vectors that will recognize only tumor-specific receptors or other surface molecules, no convincing demonstrations for such manipulations have yet been reported. Nevertheless, it does seem inevitable that these and other technical obstacles will be overcome in time, and that the in vivo delivery of tumor suppressor genes to tumors may provide one avenue to the specific control of cell proliferation in some kinds of human tumors.

References

Bartek J, Iggo R, Lane DP (1990) Genetic and immunochemical analysis of mutant p53 in human breast cancer cell lines. Oncogene 5:893–899

Bishop JM (1991) Molecular themes in oncogenesis. Cell 64:235–248

Bookstein R, Shew J-Y, Chen P-L, Scully P, Lee W-H (1990) Suppression of tumorigenicity of human prostate carcinoma cells by replacing a mutated Rb gene. Science 247:712–715

Cavanee WK, Dryja TP, Phillips RA, Benedict WF, Godbout R, Gallie BL, Murphree AL, Strong LC, White RL (1983) Expression of recessive alleles by chromosome mechanisms in retinoblastoma. Nature 305:779–784
Chang T, Yee J-K, Yeargin T, Friedman T, Haas M (1992) Suppression of acute lymphoblastic leukemia by the human p53 gene. Cancer Res 52:222–226
Chen, P-L, Chen Y, Bookstein R, Lee W-H (1990) Genetic mechanisms of tumor suppression by the human p53 gene. Science 250:1576–1580
Cheng J, Haas M (1990) Frequent mutations in the p53 tumor suppressor gene in human leukemia T-cell lines. Mol Cell Biol 10:5502–5509
Fearon ER, Vogelstein B (1990) A genetic model for colorectal carcinogenesis. Cell 61:759–767
Harris H (1988) The analysis of malignancy by cell fusion: the position in 1988. Cancer Res 48:3302–3306
Haung H-J S, Yee J-K, Shew J-Y, Chen PI-L, Bookstein R, Friedmann T, Lee EY, HP, Lee W-H (1988) Suppression of the neoplastic phenotype by replacement of the RB gene in human cancer cells. Science 242:1563–1566
Hsu IC, Metcalf RA, Sun T, Welsh JA, Wang NJ, Harris CC (1991) Mutational hotspot in the p53 gene in human hepatocellular carcinomas. Nature 350:427–428
Klein G (1987) The approaching era of the tumor suppressor genes. Science 238:1539–1545
Knudson AG Jr (1971) Mutation and cancer: statistical study of retinoblastoma. Proc Natl Acad Sci USA 68:820–823
Lane DP, Benchimol S (1990) p53: oncogene or anti-oncogene? Genes Dev 4:1–8
Levine AJ (1990) The p53 protein and its interactions with the oncogene products of the small DNA tumor viruses. Virology 177:419–426
Malkin D, Li FP, Strong LC, Fraumeni Jr JF, Nelson CE, qKim DH, Kassel J, Gryka MA, Bischoff FZ, Tainsky MA, Friend SH (1990) Germ line p53 mutations in a familial syndrome of breast cancer, sarcomas, and other neoplasms. Science 250:1233–1238
Marshall CJ (1991) Tumor suppressor genes. Cell 64:313–326
Mercer WE, Shields MT, Amin M, Sauve GJ, Appella E, Romano JW, Ullrich SJ (1990) Negative growth regulation in a gliobastoma tumor cell line that conditionally expresses human wild-type p53. Proc Natl Acad Sci USA 87:6166–6170
Nigro JM, Baker SJ, Preisinger AC, Jessup JM, Hostetter R, Cleary K, Bigner SH, Davidson N, Baylin S, Devilee P, Glover T, Collins FS, Weston A, Modali R, Harris CC, Vogelstein B (1989) Mutations in the p53 gene occur in diverse human tumor types. Nature 342:705–708
Stanbridge EJ (1990). Human tumor suppressor genes. Ann Rev Genet 24:615–657

Phenotypic Plasticity and Neural Grafting

P. H. Patterson

Summary

In repairing the damaged or diseased nervous system, it is usually assumed that grafted neurons must match the phenotype of the neurons they are meant to replace in the host. For this purpose, fetal neurons of the same type or suitable neuronal cell lines are often used. Two other possibilities are discussed here. First, adrenal chromaffin cells are a source of multipotential progenitor cells capable of generating a variety of neuronal phenotypes in culture that could be used in various types of autografts. A second, related possibility is to graft appropriately plastic neurons, and allow the host environment to provide the instructive cues necessary to promote the differentiation of phenotypes suitable for the host circuitry. If the latter paradigm proves efficacious, such grafts could serve as bioassays for instructive factors *in situ*.

Introduction

While neural grafts can ameliorate behavioral deficits by simply acting as depots for spontaneous transmitter release, the graft should fit into the host circuitry for optimal function. To be capable of sensing appropriate stimuli and stimulating or inhibiting the appropriate downstream cells, the graft must make appropriate connections with the host. To do this, grafted neurons should express receptors appropriate for the transmitter(s) of their new presynaptic partners, and they should release transmitter(s) appropriate for their new postsynaptic partners. Normally, therefore, the neurons to be grafted would be identicial to those that the graft is intended to replace. In current practice, this involves grafting the appropriate fetal tissue or neuronal cell line. The drawbacks of these approaches are:

1. the correct cell line may not be available,
2. the primary tissue may be difficult to obtain,
3. the immunosuppression necessary to prevent graft rejection may cause unacceptable problems for the patient (Sloan et al. 1991), or
4. legal or ethical considerations may prohibit use of fetal tissue (Hoffer and Olson 1991).

A straightforward approach to circumvent ethical and legal issues, as well as to obviate the need for immunosuppression, is to use autografts. Difficulties with the use of the patient's own tissue are:

1. there may not be any "extra" neurons of the appropriate type that could be used without causing additional problems, and
2. if the patient is an adult, his/her neurons may not grow as readily as comparable cells from a fetus.

To circumvent these problems, we are investigating the possibility of utilizing adrenal medullary chromaffin cells as a source of multipotential progenitor cells than can, in the presence of suitable instructive signals, be induced to express a variety of different neuronal phenotypes (Mahanthappa et al. 1990; Patterson 1990). In addition, we are interested in the possibility that the phenotypic plasticity inherent in many types of neurons culd be used to help integrate them into the host circuitry. That is, could the host environment provide the instructive cues necessary to stimulate grafted neurons to express the transmitters and receptors appropriate for that location in the system? A related question is, can grafted neurons be used to assay for instructive factors in situ?

Neuronal phenotypic plasticity in vivo

There is considerable evidence that many aspects of neuronal phenotype can be influenced by environmental cues (Patterson and Landis 1992). For instance, axonal contact with particular target tissues can alter the shape of a neuron's dendritic arbor in specific ways, as well as the type of synaptic contacts the neuron receives (Loer and Kristan 1989). In a similar vein, the particular cells that peripheral processes of sensory neurons contact can specify the central projections of those sensory neurons (Smith and Frank 1987). Moreover, target manipulation in the periphery can induce major reorganization of somatosensory maps in the adult primate cortex (cf. Pons et al. 1991). In addition, the neurotransmitter and neuropeptide phenotype can be specified by the identity of the target that the innervating axons contact. Cutaneous nerves in the adult rat that normally contain substance P lose this peptide when made to innervate muscle; conversely, when muscle afferents innervate skin, they express substance P (McMahon and Gibson 1987). During normal development *in vivo*, sympathetic neurons alter their transmitter and peptide phenotype upon contact with certain target cells, such as the sweat glands of the rat footpad (Landis 1990). Transplantation studies have further demonstrated that the sweat glands can cause this conversion in sympathetic neurons that would not normally innervate this target (Schotzinger and Landis 1988, 1990).

In addition to target tissue influences, transmitter and peptide phenotype can be controlled by the position of neuronal cell bodies (Copenhaver and Taghert 1990; Daikoku-Ishido et al. 1990), presynaptic inputs and activity (Agoston et al. 1991; Araki and Tokunaga 1990; Baker 1990; Walicke et al. 1977), as well as hormones (Chao and McEwen 1991; Fukada 1980; Steel et al. 1989; Swanson and Simmons 1989).

Identified instructive factors

Cell culture studies have demonstrated that a variety of tissue extracts and conditioned media contain factors that can regulate a number of aspects of neuronal phenotype. These properties include the number and size of dendrites (Johnson et al. 1989; Lein and Higgins 1989; Rousselet et al. 1990), transmitter receptors (Smith and kessler 1990; Halvorsen et al. 1991), electrophysiological properties (Nishi and Willard 1988), and transmitter and peptide expression (cf. Patterson 1992). A few of these factors have been purified, cloned and sequenced. These include ciliary neurotrophic factor (CNTF; Lin et al. 1989; Stockli et al. 1989) and cholinergic differentiation factor (CDF; Yamamori et al. 1989), also known as leukemia inhibitory factor (LIF; Gearing et al. 1987). CDF/LIF and CNTF up- and down-regulate the same set of transmitter and peptide genes in cultured sympathetic neurons (Saadat et al. 1989; Nawa et al. 1990; Rao et al. 1992b). Both factors can also enhance motor neuron survival (Arakawa et al. 1990; Sendtner et al. 1990; Martinou et al. 1992; Michikawa et al. 1992). These similarities in action prompted a study of their amino acid sequences that led to predictions of tertiary structural themes in common with those predicted for IL-6 (Bazan 1991). Moreover, the receptors for CNTF and CDF/LIF share homology with the signal transducing and ligand binding subunits, respectively, of the IL-6 receptor (Davis et al. 1991; Gearing et al. 1991). These relationships are further strengthened by the fact that CDF/LIF and IL-6 have identical effects on the proliferation and differentiation of several types of nonneural cells (Patterson 1992).

Other effects of CNTF include enhancing the survival of ciliary ganglion neurons *in vitro* (Barbin et al. 1984) but not *in vivo* (Oppenheim et al. 1991), and promoting the differentiation of O-2A progenitor cells to type-2 astrocytes in culture (Lillien and Raff 1990a, b). CDF/LIF enhances the survival of cultured sensory neurons (Murphy et al. 1991).

CNTF mRNA has not been detected in the rat embryo, but it is concentrated in Schwann cells and astrocytes of postnatal brain and peripheral nerves (Stockli et al. 1989, 1991). These findings suggest that this protein is not a target-derived trophic/instructive differentiation factor, but rather it may serve as a "lesion factor" for injured neurons. Levels of CDF/LIF mRNA, on the other hand, are high in postnatal rat footpads (Yamamori 1991), which contain the sweat glands that convert sympathetic neurons from the noradrenergic to the cholinergic phenotype. A protein resembling CDF/LIF has not yet been detected in soluble extracts of footpads, however (Rao and Landis 1990; Rohrer 1992; Rao et al. 1992a). Using RT-PCR and RNase protection methods, we have also recently detected CDF/LIF mRNA in neonatal rat gut, heart, skeletal muscle, spleen and thymus, as well as in adult brain (Patterson and Fann, 1992). The developmental modulation of transcript levels in these tissues suggests a role for CDF/LIF in both neuro- and hematopoietic systems.

Phenotypic "matching" of neural grafts to host sites

In addition to the cross-innervation and grafting studies with peripheral tissues mentioned above, several reports have described phenotypic plasticity in CNS grafts. Moving pieces of late fetal neocortex to heterotopic positions reveals that the differentiation of area-specific properties depends on the host site (cf., O'Leary 1989). That is, heterotopic transplants establish input and output connections appropriate for their new location. Similar conclusions have been reached by a set of experiments wherein cortical neurons are confronted with altered afferent projections. For instance, grafts of embryonic visual cortex can be organized into barrefields typical of somatosensory cortex when innervated by sensory fibers from the whisker pads (Schlaggar and O'Leary 1991). Similarly, TH-positive glomeruli reminiscent of those found in the olfactory bulb can be induced in forebrain upon removal of the presumptive olfactory bulb and innervation by olfactory sensory axons (Graziadei and Monti-Graziadei 1986; Guthrie and Leon 1989). Providing new, somatosensory, targets for retinal afferents causes the retinal synapses to differentiate with a morphology more characteristic of the somatosensensory than of the visual system (Campbell and Frost 1987).

As expected, the more immature the tissue used for grafting, the more plasticity is observed. Neuronal stem cells would therefore be a logical choice for transplantation, and expanding the numbers of cells in culture could provide the opportunity to obtain the amount of tissue necessary for grafting. This approach has not been utilized extensively as yet (Brundin et al. 1988; Strecker et al. 1989). An intriguing alternative is the use of immortalized neuronal stem cell lines. One line, HiB5, produced with a conditional oncogene so that it ceases mitosis at the body temperature of the host, displays the ability to become integrated into the granule cell layer of the hippocampus or into the cerebellum. The grafted cells can extend axons to areas appropriate for at least the hippocampal site (Renfranz et al. 1991).

Adrenal chromaffin cells as multipotential progenitor cells

Another source of multipotential neuronal progenitor cells is the adrenal medulla, whose cells could also be provided to patients as autografts. Adrenal chromaffin cells resemble in many respects the natural progenitor cells for the sympathetic and enteric nervous systems (Anderson et al. 1991; Carnahan and Patterson 1991a, b; Carnahan et al. 1991). Chromaffin cells can be converted into neurons of several different phenotypes upon the addition of appropriate trophic (fibroblast growth factor [FGF] and nerve growth factor [NGF]) and instructive factors (CDF/LIF, Doupe et al. 1985; Stemple et al. 1988), forming functional synapses upon one another in culture (Ogawa et al. 1984). Moreover, the numbers of these progenitor cells can be increased considerably by growth in medium containing FGF and dexamethasone (Mahanthappa et al. 1990).

How do chromaffin cells and their derivatives perform when grafted into the CNS? Noradrenergic axons from grafted sympathetic ganglia can innervate a denervated adult hippocampus, and the distribution of these axons resembles both that of the normal noradrenergic innervation originating from the locus coeruleus and that of the septal cholinergic afferents (Bjorklund and Stenevi 1977; Gage et al. 1984). In fact, sympathetic neurons induced to acquire cholinergic properties in culture can also survive and grow in the denervated hippocampus of adult rats (Mahanthappa et al. 1990). The results from grafting adrenal tissue are mixed, however (Fine 1990). Increased dopamine levels and transient behavioral effects have been observed after adrenal grafts near areas that normally receive dopaminergic innervation (Freed et al. 1981; Stromberg et al. 1985:; Bing et al. 1988; Bankiewicz et al. 1988; Drucker-Colin et al. 1990). It has also been reported that some chromaffin cells can sprout dopaminergic fibers when grafted into the denervated caudate nucleus (Nishino et al. 1990). However, serum and CSF catecholamine levels and the behavioral changes are not correlated with graft survival (Bankiewicz et al. 1990; Becker et al. 1990; but see also Rioux et al. 1991). On the other hand, CSF catecholamines have been correlated with decreased pain sensitivity (Sagen et al. 1991).

In fact, survival of adrenal allo- and autografts in the brain is generally quite poor (Bohn et al. 1987; Dohan et al. 1988; Hansen et al. 1988; Jankovic et al. 1989; Kordower et al. 1991), a finding not surprising in light of the known chromaffin cell requirement for high levels of corticosteroid for survival (Doupe et al. 1985). If adrenal grafts are supplemented with NGF infusion or glial cells, however, chromaffin cell survival and the reduction in apomorphine-induced rotations are significantly enhanced (Stromberg et al. 1985; Bing et al. 1990; Date et al. 1990; Hansen et al. 1990; Watts et al. 1990). This approach is also being pursued in human studies (Olson et al. 1991). We also find that survival of adrenal chromaffin cells in the adult rat striatum is greatly enhanced by cografting NGF-secreting fibroblasts, as is their expression of a variety of neuronal markers (Patterson, et al 1992). Similar results are obtained by cografting adrenal cells with nonsecreting astrocytes (Cunningham et al. 1992). The axons of such adrenal-derived neurons do not, however, leave the graft and innervate the host striatum. This could be due to much higher levels of NGF within the graft compared to levels in the striatum. NGF levels in the striatum are low relative to other areas of the CNS (Korsching et al. 1985), and NGF-sensitive axons do not grow away from areas of elevated NGF concentrations (Campenot 1979). Adrenal-derived neurons may be expected to innervate NGF-rich areas such as the hippocampus more readily, and we found that such neurons, of the cholinergic phenotype, do survive in that region (Mahanthappa et al. 1990). Whether adrenal-derived neurons will be integrated into the host circuitry remains to be determined.

Multipotential progenitors as probes for differentiation factors in situ

Primary multipotential progenitor cells or progenitor cell lines have another potential use as grafts in the CNS. Quite apart from their possible role in repair of functional disorders, such grafts could serve as in situ reporters of local neuronal differentiation factors. For instance, the sympathoadrenal progenitors and their progeny express many different neurotransmitter enzymes and neuropeptides (Lundberg and Hokfelt 1986; Morris and Gibbons 1989; Nawa and Sah 1990; Vandenbergh et al. 1991; Patterson and Landis 1992). Growth of these cells in the presence of appropriate differentiation signals up- and down-regulates these transmitter and peptide genes (Jonakait and Schotland 1990; Nawa and Patterson 1990; Speigel et al. 1990; Barbany et al. 1991; Vandenbergh et al. 1991; Patterson 1992). Furthermore, since these neurons display at least some of the same plasticity *in vivo* as they do *in vitro*, they should provide very sensitive biological probes for local differentiation signals in the host brain. One potential example of phenotypic alteration in grafted cells is the loss of tyrosine hydroxylase in PC12 cells transplanted to the adult rat brain (Freed et al. 1986; Cunningham et al. 1991).

Perspectives

Although progenitor cells can display impressive flexibility (single progenitors can give rise to most of the neuronal and glial cell types in the retina, for example; Turner et al. 1990), it would not be surprising if there are limits to this phenotypic plasticity. Nonetheless, the characteristic traits that are used to identify specific neuronal populations in particular parts of the nervous system are often expressed in other contexts as well. The various transmitters, neuropeptides and their receptors are employed across the breadth of the nervous system. Trophic factors such as NGF, once thought to be used selectively by discrete PNS populations, are now known to play similar roles in the CNS as well. Thus, progenitors from various regions of the nervous system may well share the ability to express many of the same sets of genes. On the other hand, certain monoclonal antibodies have identified neuronal antigens with quite limited distributions, and certain morphologies are unique to single types of neuron (e.g., Purkinje cells).

Another phenotypic trait that is essential for incorporation of neural grafts into the host circuitry is axon guidance to suitable target areas and recognition of the appropriate target cells. Although little is known at the molecular level about this connectivity problem, it is clear that both general guidance pathways (such as laminin) and positional information cues (e.g., Purves et al. 1981; Stahl et al. 1990; Trisler 1990) are used by each class of neuron to form its distinctive sets of synapses. Could a multipotential progenitor cell recognize these cues and form the connections appropriate for that site? Here again, the limited information available suggests that common molecules and mechanisms may be shared in various parts of the nervous system. There is evidence, for instance, that positional information required for certain interneuronal synapses may be shared with neuromuscular systems (Wigston and Sanes 1985; Suzue et al. 1990). Also relevant here is the previously cited evidence that certain of the connections made by, or on, a neuron can be specified by the other connections made by the neuron. That is, the particular target cells that a neuron contacts in the periphery can specify which cells that neuron contacts, and is contacted by, in its central projections. Moreover, grafting studies have shown that progenitor cells can assume the connectivity of the host site in which they are implanted (cf. O'Leary 1989; McConnell 1990; Wictorin et al. 1990). In sum, these results indicate that connectivity, like many other aspects of neuronal phenotype, is plastic and influenced by the environment in which the neuron develops.

Acknowledgments. The work described from this laboratory was supported by a Javits Neuroscience Investigator Award from the National Institute for Neurological Disease and Stroke, and a Neuroscience Research Project Award from the McKnight Foundation. I thank Kenji Niijima for his comments on the manuscript.

References

Agoston DV, Eiden LE, Brenneman DE (1991) Calcium-dependent regulation of the enkephalin phenotype by neuronal activity during early ontogeny. J Neurosci Res 28:140–148

Anderson DJ, Carnahan J, Michelsohn A, Patterson PH (1991) Antibody markers identify a common progenitor to sympathetic neurons and chromaffin cells in vivo, and reveal the timing of commitment to neuronal differentiation in the sympathoadrenal lineage. J Neurosci, 11:3507–3519

Arakawa Y, Sendtner M, Thoenen H (1990) Survival effect of ciliary neurotrophic factor (CNTF) on chick embryonic motoneurons in culture: Comparison with other neurotrophic factors and cytokines. J Neurosci 10:3507–3515

Araki T, Tokuaga F (1990) Norepinephrine suppresses both photoreceptor and neuron-like properties expressed by cultured rat pineal glands. Cell Differen Devel 31:129–135

Baker H (1990) Unilateral, neonatal olfactory deprivation alters tyrosine hydroxylase expression but not araomatic amino acid decarboxylase or GABA immunoreactivity. Neurosci 36:761–771

Bankiewicz KS, Plunkett RJ, Kopin IJ, Jacobowitz DM, London WT, Oldfield EH (1988) Transient behavioral recovery in hemiparkinsonian primates after adrenal medullary allografts. Progr Brain Res 78:543–549

Bankiewicz KS, Plunkett RJ, Mefford I, Kopin IJ, Olfield EH (1990) Behavioral recovery from MPTP-induced parkinsonism in monkey after intracerebral tissue implants is not related to CSF concentrations of dopamine metabolites. Progr Brain Res 82:561–571

Barbany G, Friedman WI, Persson H (1991) Lymphocyte-mediated regulation of neurotransmitter gene expression in rat sympathetic ganglia. J Neuroimmunol 32:97–104

Barbin G, Manthorpe M, Varon S (1984) Purification of the chick eye ciliary neuronotrophic factor. J Neurochem 43:1468–1478

Bazan JF (1991) Neuropoietic cytokines in the hematopoietic fold. Neuron 7:197–208

Becker JB, Curran EJ, Freed WJ, Poltorak M (1990) Mechanisms of action of adrenal medulla grafts: the possible role of peripheral and central dopamine systems. Prog Brain Res 82:499–507

Bing G, Notter MFD, Hansen JT, Gash DM (1988) Comparison of adrenal medullary, carotid body and PC12 cell grafts in 6-OHDA lesioned rats. Brain Res Bull 20:399–406

Bing G, Notter MFD, Hansen JT, Kellogg C, Kordower JH, Gash DM (1990) Cografts of adrenal medulla with C6 glioma cells in rats with 6-hydroxydopamine-induced lesions. Neuroscience 34:687–697

Bjorklund A, Stenevi U (1977) Experimental reinnervation of the rat hippocampus by grafted sympathetic ganglia. I. Axonal regeneration along the hippocampal fimbria. Brain Res 138:259–270

Bohn MC, Cupit L, Marciano F, Gash DM (1987) Adrenal medulla grafts enhance recovery of striatal dopaminergic fibers. Science 237:913–916

Brundin P, Barbin G, Strecker RE, Isacson O, Prochianz A, Bjorklund A (1988) Survival and function of dissociated dopamine neurons grafted at different developmental stages or after being cultured in vitro. Dev Brain Res 39:233–243

Campbell G, Frost DO (1987) Target-controlled differentiation of axon terminals and synaptic organization. Proc Natl Acad Sci USA 84:6929–6933

Campenot RB (1979) Development of sympathetic neurons in compartmentalized cultures. I. Local control of neurite growth by nerve growth factor. Dev Biol 93:1–12

Carnahan J, Patterson PH (1991 a) The generation of monoclonal antibodies that bind preferentially to adrenal chromaffin cells and the cells of embryonic sympathetic ganglia. N Neurosci 11:3493–3506

Carnahan J, Patterson PH (1991 b) Isolation of the progenitor cells of the sympathoadrenal lineage from embryonic symphatetic ganglia with the SA monoclonal antibodies. J Neurosci 11:3520–3530

Carnahan J, Anderson DJ, Patterson PH (1991) Evidence that enteric neurons may derive from the sympathoadrenal lineage. Dev Biol 148:552–561

Chao HM, McEwen BS 1991) Glucocorticoid regulation of neuropeptide mRNAs in the rat striatum. Mol Brain Res 9:307–311

Copenhaver PF, Taghert PH (1990) Neurogenesis in the insect enteric nervous system: generation of preimgratory neurons from anepithelial placode. Development 109:17–28

Cunningham LA, Short MP, Vielkind U, Breakefield XO, Bohn MC (1991) Survival and differentiation within the adult mouse striatum of grafted rat pheochromocytoma cells (PC12) genetically modified to express recombinant b-NGF. Exp Neurol 112:174–182

Cunningham LA, Hansen JT, Short MP, Bohn MC (1992) The use of genetically altered artocytes to provide nerve growth factor adrenal chromaffin cells grafted into the striatum. Brain Res 56:192–202

Daikoku-Ishido H, Okamura Y, Yanaihara N, Daikoku S (1990) Development of the hypothlamic luteinizing hormone-releasing hormone-containing neuron system in the rat: In vivo and in transplantation studies. Dev Biol 140:374–387

Date I, Felten SY, Felten DL (1990) Cografts of adrenal medulla with peripheral nerve enhance the survivability of transplanted adrenal chromaffin cells and recovery of the host nigrostriatal dopaminergic system in MPTP-treated young adult mice. Brain Res 537:33–39

Davis S, Aldrich TH, Valenzuela DM, Wong V, Furth ME, Squinto SP, Yancopoulos GD (1991) The receptor for ciliary neurotrophic factor. Science 253:59–63

Dohan FC, Robertson JT, Feler C, Schweitzer J, Hall C, Robertson JH (1988) Autopsy findings in a Parkinson's disease patient treated with adrenal medullary to caudate nucleus transplant. Soc Neurosi Abstr 14:8

Doupe AJ, Landis SC, Patterson PH (1985) Environmental influences in the development of neural crest derivatives: Glucocorticoids, growth factors, and chromaffin cell plasticity. J Neurosci 5:2119–2142

Drucker-Colin R, Garcida-Hernandez F, Mendoza-Ramirez JL, Pacheco-Cano MT, Komisaruk BR (1990) Possible mechanism of action of adrenal transplants in Parkinson's disease. Prog Brain Res 82:509–514

Fine A (1990) Transplantation of adrenal tissue into the central nervous system. Brain Res Rev 15:121–133

Freed WJ, Morishisa JM, Spoor E, Seiger A, Wyatt RJ (1981) Transplanted adrenal chromaffin cells in rat brain reduce lesion-induced rotational behavior. Nature 292:351–352

Freed WJ, Patel-Vaidya U, Geller HM (1986) Properties of PC12 pheochromocytoma cells transplanted to the adult rat brain. Properties of PC12 pheochromocytoma cells transplanted to the adult rat brain Exp Brain Res 63:557–566

Fukada K (1980) Hormonal control of neurotransmitter choice in sympathetic neurone cultures. Nature 287:553–555

Gage FH, Bjorklund A, Stenevi U (1984) Denervation releases a neuronal survival factor in adult rat hippocampus. Nature 308:637–639

Gearing DP, Gough NM, King JA, Hilton DJ, Nicola NA, Simpson RJ, Nice EC, Kelso A, Metcalf D (1987) Molecular cloning and expression of cDNA encoding a murine myeloid leukaemia inhibitory factor (LIF). EMBO J 6:3995–4002

Gearing DP, Thut CJ, VandenBos T, Gimpel SD, Delaney PB, King J, Price V, Cosman D and Beckman MP (1991) Leukemia inhibitory factor receptor is structurally related to the IL-6-signal transducer, gp 130. EMBO J 10:2839–2848
Gough NM, Williams RL (1989) The pleiotropic actions of leukemia inhibitory factor. Cancer Cells 1:77–80
Graziadei PPC, Monti-Graziadei GA (1986) Neuronal changes in the forebrain of mice following penetration by regenerating olfactory axons. J Comp Neurol 247:344–356
Guthrie KM, Leon M (1989) Induction of tyrosine hydroxylase expression in rat forebrain neurons. Brain Res 497:117–131
Halvorsen SW, Schid HA, AE McEachern, Berg DK (1991) Regulation of acetylcholine receptors on chick ciliary ganglion neurons by components from the synaptic target tissue. J Neurosci 11:2177–2186
Hansen JT, Kordower JH, Fiandaca MS, Jiao S-S, Notter MFD, Gash DM (1988) Adrenal medullary autografts into the basal ganglia of Cebus monkeys: graft viability and fine structure. Exp Brain Res 102:65–75
Hansen JT, Fiandaca MS, Kordower JH, Notter MFD, Gash DM (1990) Striatal adrenal medulla/sural nerve cografts in hemiparkinsonian monkeys. Prog Brain Res 82:573–580
Hoffer BJ, Olson L (1991) Ethical issues in brain-cell transplantation. Trends Neurosci 14:384–388
Jancovic J (1989) Adrenal medullary autografts in patients with Parkinson's disease. New Eng J Med 321:325–326
Johnson MI, Higgins D, Ard MD (1989) Astrocytes induce dendritic development in cultured sympathetic neurons. Dev Brain Res 47:289–292
Jonakait GM, Schotland S (1990) Conditioned medium from activated splenoctyes increases substance P in sympathetic ganglia. J Neurosci Res 26:24–30
Kordower JH, Cochran E, Penn RD, Goetz CG (1991) Putative chromaffin cell survival and enhanced host-derived TH-fiber innervation following a functional adrenal medulla autograft for Parkinson's disease. Ann Neurol 29:405–412
Korsching S, Auberger G, Heumann R, Scott J, Thoenen H (1985) Levels of nerve growth factor and its mRNA in the central nervous system of the rat correlate with cholinergic innervation. EMBO J 4:1389–1393
Landis SC (1990) Target regulation of neurotransmitter phenotype. Trends Neurosci 13:344–350
Lein PJ, Higgins D (1989) Laminin and a basement membrane extract have different effects on axonal and dendritic outgrowth from embryonic rat sympathetic neurons in vitro. Dev Biol 136:330–345
Lillien LE, Raff MC (1990a) Analysis of the cell-cell interactions that control type-2 astrocyte development in vitro. Neuron 4:525–534
Lilien LE, Raff MC (1990b) Differentiation signals in the CNS: Type-2 astrocyte development in vitro as a model system. Neuron 5:111–119
Lin L-FH, Mismer D, Lile JD, Armes LG, Butler ET, Vannice JL, Collins F (1989) Purification, cloning and expression of ciliary neurotrophic factor (CNTF). Science 246:1023–1025
Loer CM, Kristan WB Jr (1989) Central synaptic inputs to identified leech neurons determined by peripheral targets. Science 244:64–66
Lundberg JM, Hokfelt T (1986) Multiple coexistence of peptides and classical neurotransmitters in peripheral autonomic and sensory neurons – functional and pharmacological implications. In: Hokfelt T, Fuxe K, Pernow B (eds.) Progress in brain research. Elsevier, Amsterdam
Mahanthappa NG, Gage FH, Patterson PH (1990) Adrenal chromaffin cells as multipotential neurons for autografts. Prog Brain Res 82:33–39
Martinou JC, Martinou I, Kato AC (1992) Cholinergic differentiation factor (CDF/LIF) promotes survival of isolated rat embryonic motoneurons in vitro. Neuron 8:737–744
McConnell SK (1990) The specification of neuronal identity in the mammalian cerebral cortex. Experientia 46:922–929
McMahon SB, Gibson S (1987) Peptide expression is altered when afferent nerves reinnervate inappropriate tissue. Neurosci Lett 73:9–15

Michikawa M, Kikuchi S, Kim SU (1992) Leukemia inhibitory factor (LIF) mediated increase of choline acetyltransferase activity in mouse spinal cord neurone in culture. Neurosci Lett 140:75–77

Morris JL, Gibbins IL (1989) Co-localization and plasticity of transmitters in peripheral autonomic and sensory neurons. Int J Devel Neurosci 7:521–531

Murphy M, Reid K, Hilton DJ, Bartlett PF (1991) Generation of sensory neurons is stimulated by leukemia inhibitory factor. Proc Natl Acad Sci USA 88:3498–3501

Nawa H, Patterson PH (1990) Separation and partical characterization of neuropeptide-inducing factors in heart cell conditioned medium. Neuron 4:269–277

Nawa H, Sah D (1990) Different biological activities in conditioned media control the expression of a variety of neuropeptides in cultured sympathetic neurons. Neuron 4:279–287

Nawa H, Yamamori T, Le T, Patterson PH (1991) The generation of neuronal diversity: Analogies and homologies with hematopoiesis. Cold Spring Harbor Symp 55:247–253

Nishi R, Willard AL (1988) Conditioned medium alters electrophysiological and transmitter-related properties expressed by rat enteric neurons in cell culture. Neuroscience 25:759–769

Nishino H, Hashitani T, Kumazaki M (1990) Phenotypic plasticity of grafted catecholaminergic cells in the dopamine-depleted caudate nucleus of the rat. Neurosci Res Suppl 13:S54–S60

Ogawa M, Ishikawa T, Immajin A (1984) Adrenal chromaffin cells form functional cholinergic synapses in culture. Nature 307:66–68

O'Leary DDM (1989) Do cortical areas emerge from a protocortex? Trends Neurosci 12:400–406

Olson L, Backlund E-O, Ebendal T, Freedman R, Hamberger B, Hansson P, Hoffer B, Lindblom U, Meyerson B, Stromberg I, Sydow O, Seiger A (1991) Intraputaminal infusion of nerve growth factor to support adrenal medullary autografts in Parkinson's disease. Arch Neurol 48:373–381

Oppenheim RW, Prevette D, Qin-Wei Y, Collins F, McDonald J (1991) Control of embryonic motoneuron survival in vivo by ciliary neurotrophic factor. Science 251:1616–1618

Patterson PH (1990) Control of cell fate in a vertebrate neurogenic lineage. Cell 62:1035–1038

Patterson PH, Fann M-J (1992) The tissue distribution of CDF/LIF mRNA and its implications for neuro- and hematopoiesis. In: CIBA Foundation Symposium 167:125–140

Patterson PH, Landis SC (1992) Phenotype specifying factors and the control of neuronal differentiation decisions. In: Hendry I, Hill C (eds.) Development, regeneration and plasticity of the autonomic nervous system. Elsevier, London, pp 231–265

Patterson PH, Niijima K, Chalmero GA, Peterson DA, Fisher LT, Gage FH (1992) Adrenal chromaffin cell co-grafted with NGF-secreting fibroblasts. Restorative Neurol Neurosci 4:2704

Patterson PH (1992) The emerging neuropoietic cytokine family: first CDF/LIF, CNTF and IL-6, next ONC, MGF and GCSF? Curr Opinion Neurobiol 2:94–97

Patterson PH (1992) Instructive neuronal differentiation factors. In: Loughlin S, Fallon J (eds.) Neurotrophic factors. Academic Press, New York, pp 527–563

Pons TP, Garraghty PE, Ommaya AK, Kaas JH, Taub E, Mishkin M (1991) Massive cortical reorganization after sensory deafferentation in adult Macaques. Science 252:1857–1860

Purves D, Thompson W, Yip JW (1981) Reinnervation of ganglia transplanted to the neck from different levels of the guinea-pig sympathetic chain. J Physiol 313:49–63

Rao MS, Landis SC (1990) Characterization of a cholinergic neuronal differentiation factor in sweat glands. Neuron 5:899–910

Rao MS, Patterson PH, Landis SC (1992a) Multiple cholinergic differentiation factors are present in footpad extracts: Comparison with known Cholinergic factors. Development, in press

Rao MS, Tyrrell S, Landis SC, Patterson PH (1992b) Effects of ciliary neurotrophic factor (CNTF) and depolarization on neuropeptide expression in cultured sympathetic neurons. Dev Biol 150:281–293

Renfranz PJ, Cunningham MG, McKay RDG (1991) Region-specific differentiation of the hippocampal stem cell line HiB5 upon implantation into the developing mammalian brain. Cell 66:713–729

Rioux L, Gaudin DP, Bui LK, Gregoire L, DiPaolo T, Bedard PJ (1991) Correlation of functional recovery after a 6-hydroxydopamine lesion with survival of grafted fetal neruons and release of dopamine in the striatum of the rat. Neurosci 40:123–131

Rohrer H (1992) Cholinergic neuronal differentiation factors: evidence for the presence of both CNTF-like and non-CNTF-like factors in developing cut footpad. Development 114:689-698

Rousselet A, Autillo-Touati A, Araud D, Prochiantz A (1990) In vitro regulation of neuronal morphogenesis and polarity by astrocyte-derived factors. Dev Biol 137:33–45

Saadat S, Sendtner M, Rohrer H (1989) Ciliary neurotrophic factor induces cholinergic differentiation of rat sympathetic neurons in culture. J Cell Biol 108:1807–1816

Sagen J, Kemmler JE, Wang H (1991) Adrenal medullary transplants increase spinal cord cerebrospinal fluid catecholamine levels and reduce pain sensitivity. J Neurochem 56:623–627

Schlagger BL, O'Leary DDM (1991) Potential of visual cortex to develop an array of functional units unique to somatosensory cortex. Science 252:1556–1560

Schotzinger RJ, Landis SC (1988) Cholinergic phenotype developed by noradrenergic sympathetic neurons after innervation of a novel target in vivo. Nature 335:637–639

Schotzinger RJ, Landis SC (1990) Acquisition of cholinergic and peptidergic properties by sympathetic innervation of rat sweat glands requires interaction with normal target. Neuron 5:91–100

Sendtner M, Kreutzberg GW, Thoenen H (1990) Ciliary neurotrophic factor prevents the degeneration of motor neurons after axotomy. Nature 345:440–441

Sloan DJ, Wood MJ, Charlton HM (1991) The immune response to intracerebral grafts. Trends Neurosci 14:341–346

Smith CL, Frank E (1987) Peripheral specification of sensory neurons transplanted to novel locations along the neuraxis. J Neurosci 7:1537–1549

Smith KE, Kessler JA (1990) Non-neuronal cell-conditioned medium regulates muscurinic receptor expression in cultured sympathetic neurons. J Neurosci 8:2406–2413

Speigel K, Wong V, Kessler JA (1990) Translational regulation of somatostatin in cultured sympathetic neurons. Neuron 4:303–411

Stahl B, Muller B, von Boxberg Y, Cox EC, Bonhoeffer F (1990) Biochemical characterization of a putative axonal guidance molecule of the chick visual system. Neuron 5:735–74

Steel JH, Gon G, O'Halloran DJ, Jones PM, Yanaihara N, Ishikawa H, Bloom SR, Polak JM (1989) Galanin and vasoactive intestinal polypeptide are colocalized with classical pituitary hormones and show plasticity of expression. Histochemistry 93:183–189

Stemple DL, Mahanthappa NK, Anderson DJ (1988) Basis FGF induces neuronal differentiation, cell division, and NGF dependence in chromaffin cells: A sequence of events in sympathetic development Neuron 1:517–525

Stockli KA, Lottspeich F, Sendtner M, Masiakowski P, Carroll P, Lindholm D, Thoenen H (1989) Molecular cloning, expression and regional distribution of rat ciliary neurotrophic factor. Nature 342:920–923

Stockli KA, Lilien LE, Naher-Noe M, Breitfeld G, Hughes RA, Raff MC, Thoenen H, Sendtner M (1991) Regional distribution, developmental changes and cellular localization of CNTF-mRNA and protein in the rat brain. J Cell Biol, in press

Strecker R, Miao R, Loring J (1989) Survival and function of aggregate cultures of rat fetal dopamine neurons grafted in a rat model of Parkinson's disease. Exp Brain Res 76:315–322

Stromberg I, Herrera-Marschitz M, Ungerstedt U, Ebendahl T, Olson L (1985) Chronic implants of chromaffin tissue into the dopamine-denervated striatum. Effects of NGF on graft survival, fiber outgrowth and rotational behavior. Exp Brain Res 60:335–349

Suzue T, Kaprielian Z, Patterson PH (1990) A monoclonal antibody that defines rostrocaudal gradients in the mammalian nervous system. Neuron 5:421–431

Swanson LW, Simmons DM (1989) Differential steroid hormone and neural influences on peptide mRNA levels in CRH cells of the paraventricular nucleus: A hybridization histochemical study in the rat. J Comp Neurol 285:413–435

Trisler D (1990) Cell recognition and pattern formation in the developing nervous system. J Exp Biol 153:11–27

Turner DL, Snyder EY, Cepko CL (1990) Lineage-independent determination of cell type in the embryonic mouse retina. Neuron 4:833–845

Vandenbergh DJ, Mori N, Anderson DJ (1991) Co-expression of multiple neurotransmitter enzyme genes in normal and immortalized sympathoadrenal progenitor cells. Dev Biol 148:10–22

Walicke PA, Campenot RB, Patterson PH (1977) Determination of transmitter function by neuronal activity. Proc Natl Acad Sci USA 74:5767–5771

Watts RL, Bakay RAE, Herring CJ, Sweeney KM, Colbassani HJ, Mandir A, Byrd LD, Iuvone PM (1990) Preliminary report on adrenal medullary grafting and cografting with sural nerve in the treatment of hemiparkinson monkeys. Progr Brain Res 82:581–591

Wictorin K, Brundin P, Gustavil B, Lindvall O, Bjorklund A (1990) Reformation of long axon pathways in adult rat central nervous system by human forebrain neuroblasts. Nature 347:556–558

Wigston DJ, Sanes JR (1985) Selective reinnervation of intercostal muscles transplanted from different segmental levels to a common site. J Neurosci 5:1208–1221

Yamamori T, Fukada K, Aebersold R, Korsching S, Fann MJ, Patterson PH (1989) Cholinergic neuronal differentiation factor from heart cells is identical to leukemia inhibitory factor. Science 246:1412–1416

Yamamori T (1991) Localization of CDF/LIF mRNA in the rat brain and peripheral tissues. PNAS 88:7298–7302

The Sympathoadrenal Progenitor of the Neural Crest: Basic Biology and Therapeutic Potential

D. J. Anderson

Summary

The development of successful cell-replacement therapies in the nervous system would be enhanced by an understanding of the basic biology of embryonic neural progenitor cells. The sympathoadrenal (SA) progenitor of the peripheral nervous system is among the best-characterized of such progenitors, in vertebrates. This cell, which derives from the neural crest, has a restricted developmental program that allows it to develop into a chromaffin cell or a sympathetic neuron, depending upon the environment to which it migrates. The extracellular signals that control the fate of SA progenitors have been identified, using both primary cultures of cells isolated from fetal rat adrenal glands and immortalized cell lines (MAH cells) derived from these primary progenitors. These studies have suggested that glucocorticoid hormones secreted by the adrenal cortex promote the development of SA progenitors into chromaffin cells, whereas fibroblast growth factor (FGF) and nerve growth factor (NGF) promote their development into sympathetic neurons. FGF and NGF appear to act sequentially; the former factor promotes neurite outgrowth and proliferation while the latter factor supports the survival of post-mitotic neurons. In addition to choosing between neuronal and endocrine fates, SA progenitors must also express an appropriate neurotransmitter phenotype. Although ordinarily noradrenergic, sympathetic neurons are able to convert to a cholinergic phenotype in response to specific inducing factors. This cholinergic potential is already present in uncommitted SA progenitors, as evidenced by their ability to synthesize acetylcholine. Recent studies suggest that these cells may have additional developmental capacities besides the chromaffin, noradrenergic and cholinergic sympathetic phenotypes, including the ability to generate serotonergic enteric neurons. The SA progenitor may therefore be a "master" neuroendocrine progenitor for the peripheral nervous system. Because it is multipotential, this progenitor (or immortalized lines derived from it) may be useful in heterologous cell transplantation therapies for a variety of neurodegenerative diseases.

Introduction

The transplantation of cells into the nervous system is attracting increasing attention as a potential therapy for the treatment of neurological and neurodegenerative disorders (Björklund 1991). A variety of transplantation strategies have been proposed and tested, including the grafting of normal, immortalized and genetically modified cells (Gage et al. 1987). Cell transplants can perform a number of therapeutic functions. They may provide neurotransmitters to replace those normally released by degenerating neurons; they may provide neurotrophic factors to prevent the degeneration of endogenous neurons; they may protect endogenous neurons against neurotoxic substances; or, ideally, they may replace the function of degenerating cells (Gage and Buzsáki 1989). This last function requires the establishment of appropriate connectivity between the grafted neurons and their endogenous targets, a formidable objective.

A major issue in the design of transplantation strategies concern the origin of the cells to be grafted. One question, for example, is whether to use homologous or heterologous cells. Although homologous cell (i.e., those from the same region and stage of maturity of the brain) might in principle seem most desireable, access to the appropriate cell population may be severely limited, depending upon the area of the nervous system of interest. An alternative is to use heterologous cells, from more accessible sources, that exhibit some of the desired properties of the homologous cell population. For example, adrenal chromaffin cells have been used to replace degenerating striatal dopaminergic neurons due to their ability to synthesize and release catecholamines (Freed et al. 1990), as have fibroblasts genetically engineered to express the tyrosine hydroxylase (TH) gene (Gage et al. 1991).

Another question concerns the optimal age of the donor cells. It has become increasingly clear that fetal cells and tissues survive and integrate into the adult host nervous system more effectively than do their more mature counterparts (Lindvall et al. 1990). This is not unexpected, as undifferentiated cells are likely to be more plastic than differentiated cells. Such plasticity may be exploited in the use of immature heterologous donor cells that may be able to alter their differentiated phenotype in response to the environment of the graft site. For example, neonatal sympathetic neurons converted to a cholinergic phenotype by a specific differentiation factor, CDF (Yamamori et al. 1989), have been tested for their ability to replace basal forebrain cholinergic neurons in an animal model of Alzheimer's disease (Mahanthappa et al. 1990). A major limitation in the wider application of such strategies is our general lack of knowledge about the nature of neuronal progenitor cells in the vertebrate embryonic nervous system. This limitation points up the desirability of learning more about the basic biology of such cells (Cattaneo and McKay 1991). In what follows, I shall discuss new insights into the control of neuronal development in vertebrates, from studies of the best-characterized vertebrate neuronal progenitor cell. This cell is the sympathoadrenal (SA) progenitor of the peripheral nervous system (Patterson 1990), a cell that gives rise to autonomic sympathetic neurons and chromaffin cells of the adrenal medulla.

Natural history of the SA progenitor

Shortly after neural tube closure, neural crest cells become separated from the CNS neuroepithelium at the dorsal margins of the neural tube (Fig. 1). These cells then migrate to diverse sites in the embryo, where they differentiate into a variety of cell types (LeDouarin 1982). These cell types include autonomic, sensory and enteric neurons and the glia of the peripheral nervous system, as well as non-neuronal cell types such as melanocytes. Some of the neural crest cells that take a ventro-lateral migration pathway, passing through the somitic mesenchyme and near the notochord, develop into sympathetic neurons which aggregate to form a rostro-caudal chain of ganglia adjacent to the dorsal aorta (Fig. 1). Other cells continue migrating ventrally to invade the developing adrenal gland, where they form the chromaffin cells of the medulla. Still others populate the region ventral to the aorta, including the foregut, in anterior regions of the embryo. Following their localization to these various sites of migratory arrest, the cells differentiate and express recognizable phenotypes.

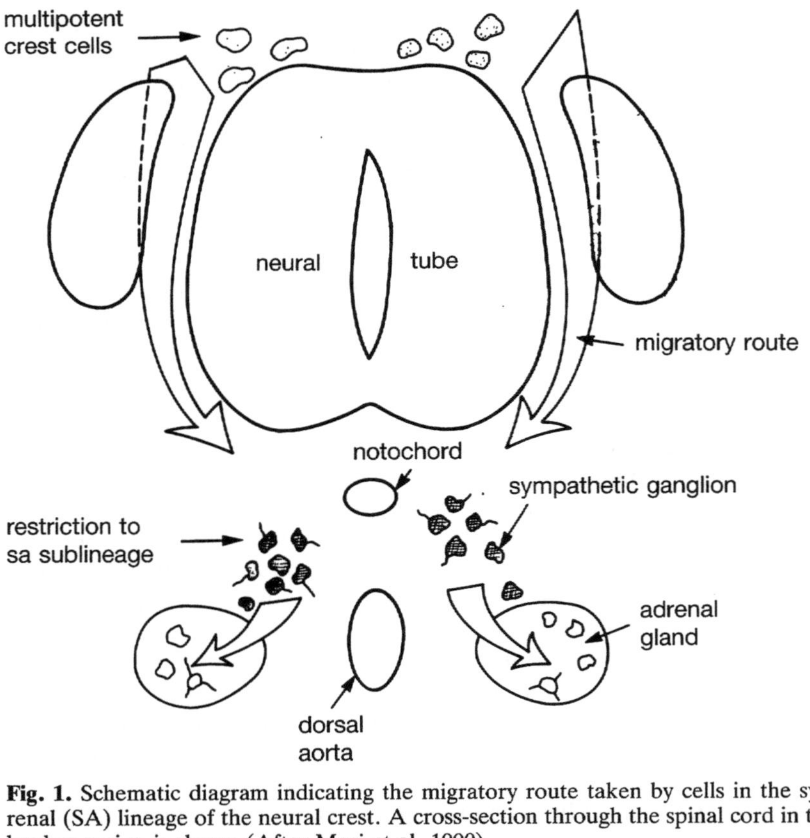

Fig. 1. Schematic diagram indicating the migratory route taken by cells in the sympathoadrenal (SA) lineage of the neural crest. A cross-section through the spinal cord in the thoraco-lumbar region is shown (After Mori et al. 1990)

The cellular analysis of development in this system has concerned two main issues:

1. what are the choices of cell fate available to individual neural crest cells at different stages of development, and
2. if the cells have choices, what environmental signals influence their fate?

In answer to the first question, it now seems clear that the earliest crest cells are multipotent, able to give rise to most or all of the derivatives that originate from the population as a whole. This finding has been established both by clonal analysis in cell culture (Baroffio et al. 1988), and by microinjection of lineage tracers into single neural crest cells *in vivo* (Bronner-Fraser and Fraser 1989). Despite this early multipotency, however, has development proceeds restrictions in developmental potential seem to occur (Anderson 1989). This has been most clearly established in the case of the SA progenitor (see below), which appears to have lost the ability to give rise to Schwann cells (D. Stemple and D. J. Anderson, unpublished) but has retained the capacity to differentiate to either a sympathetic neuron or a chromaffin cell (Anderson and Axel 1986; Carnahan and Patterson 1991b). Although SA progenitors appear to be "committed" to a specific sublineage of the neural crest, their final choice of fate is still determined, in large part, by extracellular signals located in the different environments to which they migrate. As discussed in more detail below, glucocorticoids (GC) in the adrenal gland appear to promote chromaffin cell differentiation, whereas a combination of growth factors promotes neuronal differentiation in the sympathetic ganglia. The environmental signals affecting the development of other neural crest derivatives are less well-defined.

Isolation of SA progenitors

The development of SA progenitors has been studied by isolating them from rat embryos and growing them in dissociated cell culture, where their environment can be easily manipulated. These cells are isolated using specific monoclonal antibodies to cell surface antigens, which enables the separation of progenitors from unwanted cell types by fluorescence-activated cell sorting (FACS). For example, the monoclonal antibody HNK-1, directed against a carbohydrate epitope and originally identified as a marker for a subset of human T cells, has been used to isolate SA progenitors from the fetal adrenal gland at E14.5 (Anderson and Axel 1986). A cocktail of monoclonal antibodies (SA 1–5) recently generated by immunosuppression (Carnahan and Patterson 1991a) has also been used to isolate SA progenitors from the superior cervical sympathetic ganglion, at a similar stage of development (Carnahan and Patterson 1991b).

SA progenitors grown in primary culture (Fig. 2A) have at least three fates:

1. they can develop into neurons and survive;
2. they can develop into neurons and die; or
3. they can develop into adrenal chromaffin cells.

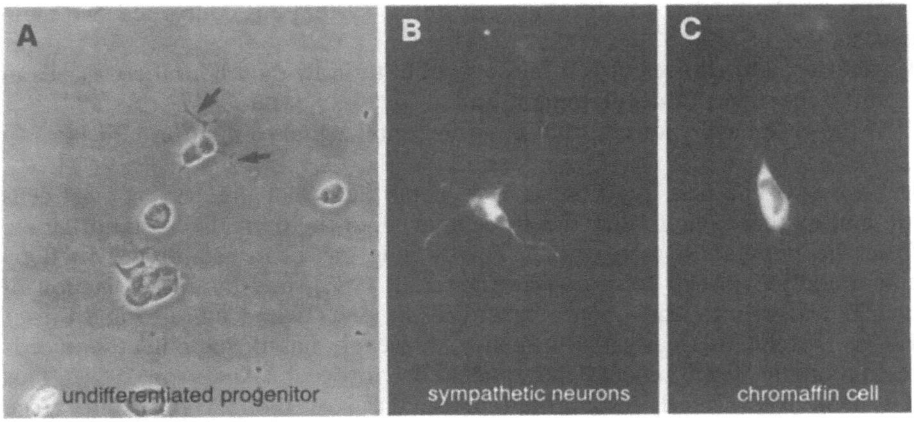

Fig. 2A–C. FACS-isolated SA progenitors develop into sympathetic neurons or chromaffin cells, in vitro. *A*, phase contrast micrographs of progenitor cells after 24 hours in culture. Arrows indicate the short cytoplasmic extensions characteristic of such cells. *B*, progenitors after 2–3 days in serum-containing medium. Under these conditions, the cells begin to differentiate into neurons, extending short neurites, but they die shortly thereafter. Neurite extension is promoted by addition of basis FGF (not shown). *C*, progenitors after 2–3 days in 5 μM dexamethasone. Note the difference in morphology from the cells in B. Cells under both conditions are labeled with an anti-TH antibody, as shown in these epifluorescence micrographs. Cells such as those in B also express neuronal markers such as neurofilament and SCG10, whereas those in C express the chromaffin marker PNMT (not shown). (After Mori et al. 1990)

The addition of exogenous GC is essential for chromaffin cell development (Fig. 2C), supporting the idea that the high local concentration of these steroids produced by the adrenal cortex controls the fate of invading SA progenitor cells *in vivo* (Doupe et al. 1985; Unsicker et al. 1978). GC play a dual role in determining the fate of SA progenitors; they act negatively, to inhibit neuronal differentiation (Unsicker et al. 1978), and positively, to promote chromaffin differentiation (Bohn et al. 1981; Seidl and Unsicker 1989). This dual action of GC is consistent with the fact that the GC receptor is a transcriptional regulatory protein that can both activate and repress the expression of different target genes (Beato 1989). Recent studies have suggested that the two effects of GC on developing chromaffin cells are temporally separated: the inhibition of neuronal differentiation can be detected several days prior to the activation of the chromaffin-specific enzyme, phenylethanolamine-N-methyltransferase) (PNMT; Michelsohn and Anderson 1992). The early effects of GC appear to be a developmental prerequisite for their later effects, since if progenitors are permitted to differentiate along the competing, neuronal pathway they rapidly lose competence to respond to GC (Anderson et al. 1991; Michelsohn and Anderson 1992). Thus, although chromaffin differentiation is controlled by a well-defined environmental signal, the response of SA progenitors to this signal is dynamic and changes with time and with direction of differentiation. None of these changes appears to be due to changes in the expression of the GC receptor, but rather to changes in the response properties of the differentiating cells (Anderson and Michelsohn 1989).

In the absence of GC, SA progenitors grown on a poly-D-lysine/laminin substrate extend neurites and initiate neuronal differentiation (Fig. 2B). This apparently spontaneous neuronal differentiation is observed in a variety of defined media, suggesting that it reflects a cell-autonomous process (S. Birren and D. J. Anderson, unpublished observations). The rate of neurite extension is greater at high cell densities, suggesting the involvement of an autocrine mechanism. This behavior of the SA progenitor suggests that the neuronal pathway of differentiation may be preferred or intrinsically favored over the chromaffin pathway. However, that rate of neurite extension can be dramatically promoted by specific growth factors (see below). In addition, the survival of sympathetic neurons is dependent upon a specific exogenous signal, nerve growth factor (NGF). Although NGF plays a key role in the development of sympathetic neurons, SA progenitors isolated from E 14.5 embryos are initially unresponsive to this factor (Anderson and Axel 1986; Birren and Anderson 1990). This observation raises the related questions of why the cells are unresponsive, and what mechanisms control the acquisition of a functional response to NGF.

Immortalization of SA progenitors reveals a growth factor relay that controls neuronal development

To address the issue of NGF-responsiveness, it became necessary to routinely generate larger numbers of cells than could be obtained by isolation of primary SA progenitors, for analysis at the molecular level. This was accomplished by immortalizing SA progenitors using a recombinant retrovirus harboring the avian *v-myc* oncogene, and establishing clonal cell lines, called MAH cells (Birren and Anderson 1990). The morphology (Fig. 3A) and antigenic phenotype of MAH cells suggests that they have retained many of the properties of their non-immortal counterparts. However, because of the possibility that immortalization may have perturbed the developmental properties of the cells, MAH cells are not used as a substitute for primary cells, but rather as a system in which to rapidly perform experiments whose outcome may then be confirmed using primary SA progenitors.

MAH cells are unresponsive to NGF by several functional criteria, including neurite outgrowth, stimulation of DNA synthesis and induction of neuron-specific mRNAs (Birren and Anderson 1990). A similar lack of NGF responsiveness is observed in primary SA progenitors (S. Birren and D. J. Anderson, unpublished observations). The lack of a response to NGF is correlated with the lack of expression of p75, the low-affinity NGF receptor (LNGFR). This has been demonstrated in MAH cells by Northern blot analysis (Birren and Anderson 1990) and in primary progenitors by immunocytochemistry using a

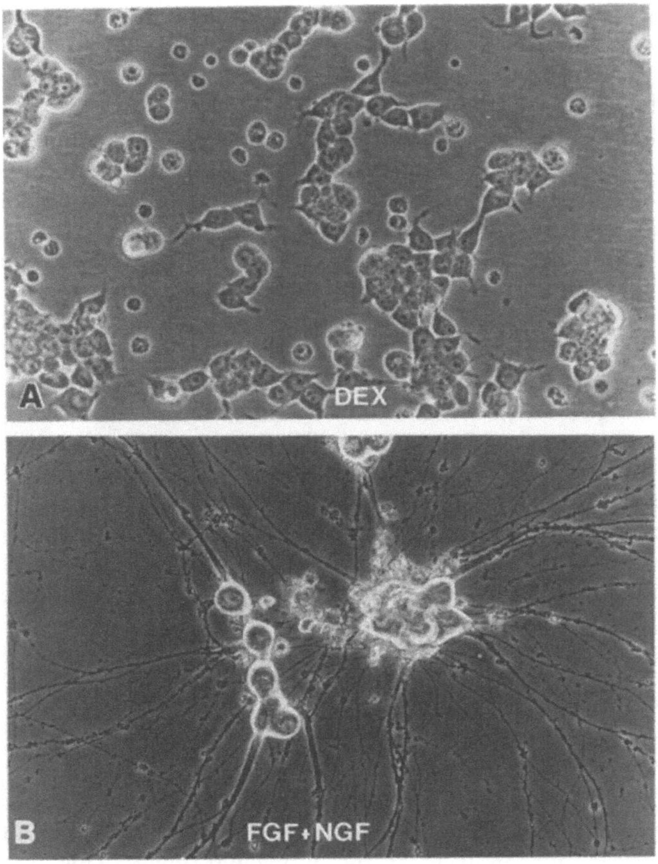

Fig. 3 A, B. Immortalized sympathoadrenal progenitor cells can differentiate into stable, postmitotic neurons. *A*, MAH cells maintained in control conditions (5 μM dexamethasone). Note the morphologic similarity to the primary cells shown in Figure 2A. *B*, a small proportion (< 0.1%) of MAH cells develop into neurons in the presence of bFGF (10 ng/ml) + NGF (50 ng/ml). The cells shown had been maintained for approximately four weeks in culture at the time this photograph was taken. The magnification in A is twice that in B. (After, Mori et al. 1990).

specific monoclonal antibody (S. Birren and D. J. Anderson, unpublished data). As primary chromaffin cells were previously shown capable of initially responding to basic fibrobast growth factor (bFGF) in a manner similar to NGF (Claude et al. 1988; Stemple et al. 1988), the effects of bFGF on MAH cells were examined. bFGF was found to induce neurite outgrowth and proliferation in MAH cells, whereas a number of other growth factors failed to elicit these responses (Birren and Anderson 1990). More recent studies have confirmed that bFGF also promotes neurite outgrowth and DNA synthesis in primary SA progenitors (S. Birren and D. J. Anderson, unpublished data). As mentioned above, however, primary progenitors show some level of neurite outgrowth in the absence of exogenous bFGF, whereas in MAH cells and in primary chromaffin cells neurite outgrowth is dependent upton bFGF. The reason for this difference is unknown.

Although bFGF acts as a neurite outgrowth-promoting factor and mitogen for both MAH cells and primary chromaffin cells, it does not support the survival of post-mitotic sympathetic neurons. NGF alone is able to provide trophic support for these neurons. In primary chromaffin cells, however, bFGF appears to induce a dependence upon NGF (Stemple et al. 1988). We therefore asked whether bFGF could influence the response of MAH cells to NGF. In the presence of both bFGF and NGF, a small proportion (~ 0.1%) of MAH cells survive and differentiate to post-mitotic sympathetic neurons (Fig. 3B). These neurons are dependent upon NGF for survival, implying that they express functional NGF receptors and that bFGF therefore induces the *the de novo* expression of the NGF receptor (Birren and Anderson 1990). Indeed, bFGF induces a low level of p75 LNGFR mRNA in MAH cells, and such an induction has also been observed in primary cells by antibody staining (S. J. Birren and D. J. Anderson, unpublished data). Taken together, these data have suggested that neuronal differentiation in this system may involve a growth factor relay, or cascade, in which bFGF (or another factor with similar biological activity) acts to promote initial neuronal differentiation, and simulatenously induces expression of the receptor for NGF, which in turn maintains the subsequent maturation and survival of the developing neurons (Fig. 4). A similar interaction between bFGF and NGF has been observed for a population of striatal neuroepithelial precursors (Cattaneo and McKay 1990). Moreover, *in situ* hybridization analyses have demonstrated that FGF receptor is expressed in the ventricular zone of the neural tube, where proliferating neural precursors are located, whereas p75 LNGFR is expressed in the marginal zone containing differentiating neurons (Heuer et al. 1990). Thus the relationship between the actions of bFGF and NGF observed for SA progenitors in the peripheral nervous system may have relevance to the differentiation of CNS neurons, as well.

Recently, a second NGF receptor has been identified, the product of the protooncogene p140 *trk* (Kaplan et al. 1991 a; Klein et al. 1991). This receptor contains a cytoplasmic tyrosine kinase domain (Kaplan et al. 1991 b; Martin-Zanca et al. 1986) and is thought to be necessary for high affinity NGF binding and signal transduction in response to NGF, perhaps as part of a heteromeric complex with p75 LNGFR (Hempstead et al. 1991). Some data suggest, however, that p140 *trk* may be sufficient for a functional response to NGF (Klein et al. 1991). Preliminary experiments indicate that MAH cells express little or no p140 *trk* mRNA (S. J. Birren, J. Verdi and D. J. Anderson, unpublished). Studies are underway to investigate the factors that control the expression of *trk* in developing SA progenitors, and the relative contributions of *trk* and p75 LNGFR to the functional response to NGF in these cells (Birren et al. 1992).

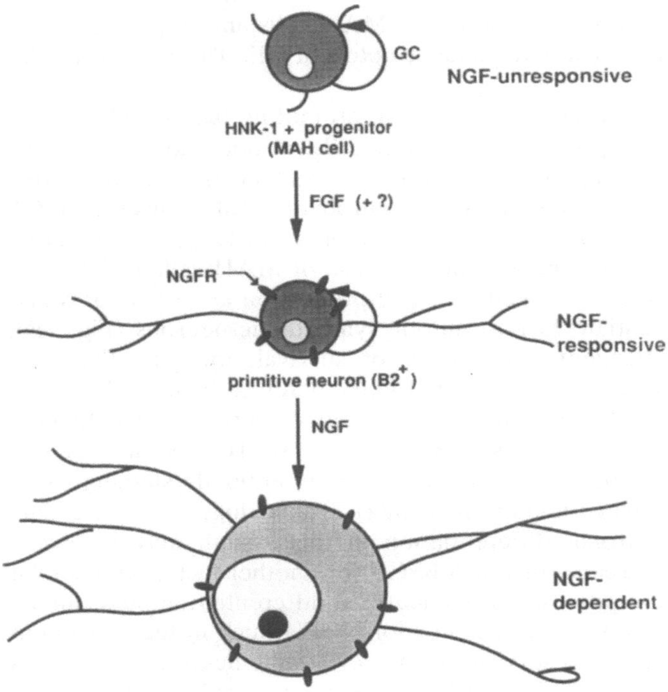

Fig. 4. Stages of neuronal differentiation in the sympathoadrenal lineage. The sympathoadrenal progenitors represented by MAH cells (*top*) extend neurites in response to FGF but do not yet express NGF receptors. In the presence of glucocorticoid (GC), these cells divide (*circular arrow*) but do not differentiate. Following treatment with FGF, a primitive neuron can be observed that bears processes and is still dividing. This cell type has been previously described in primary cultures (Anderson 1988; Anderson and Axel 1986). At least a subpopulation of these primitive neurons begins to express the NGF receptor. In the presence of NGF, these cells mature into postmitotic neurons with large cell bodies, prominent nucleoli, and extensive networks of neurites. Such neurons are then dependent on NGF for suvival. (From Birren and Anderson 1990).

Transplantation of MAH cells in a rat model of Parkinson's Disease

The use of adrenal chromaffin cell autografts for the treatment of Parkinson's disease has met with limited success (Allen et al. 1989). However, experiments in animals have indicated that fetal cells may survive better in grafts than their adult counterparts (Brundin et al. 1986). Because MAH cells are an immortal line derived from fetal chromaffin cell precursors, their ability to survive and promote functional recovery in a rat model of Parkinson's disease was tested, in collaboration with Dr. F. H. Gage (University of California, San Diego). The design of these experiments was based upon the information obtained from studies of MAH cells *in vitro* (see above). Prior to grafting, MAH cells were exposed to bFGF, and then transplanted together with RAT1 fibroblasts secreting NGF, in the hope that this condition would promote the differentiation and survival of neurons *in vivo*. Encouragingly, we found significant functional recovery (up to 50%) in animals grafted in this manner, but not in control animals receiving either untreated MAH cells or bFGF-treated MAH cells without a co-graft of NGF-secreting fibroblasts (F. Gage, S. J. Birren and D. J. Anderson, unpublished data). Unfortunately, however, this apparent functional recovery was not accompanied by the survival of large numbers of cells in the graft site. Thus, further experiments will be required to define the optimal conditions for the survival of grafted MAH cells in the CNS. Nevertheless, this system provides the opportunity to investigate cell replacement therapies using an immortalized cell line of defined embryonic origin.

The SA progenitor may be a "master" neuroendocrine progenitor for the PNS

The foregoing discussion has treated the SA progenitor as a bipotential cell, able to differentiate to a chromaffin cell in the environment of the adrenal gland and to a sympathetic neuron in the ganglionic environment. Is it possible, however, that this progenitor might have other developmental potentials, which could be elicited in different environments? If so, it would not only inform our understanding of the basic biology of this cell, but greatly expand the range of potential application in transplantation therapies.

It was already known that noradrenergic sympathetic neurons have the capacity to express a cholinergic phenotype, in response to the cholinergic differentiation factor/leukemia inhibitory factor (CDF/LIF; Fukada 1985; Patterson and Chun 1977; Yamamori et al. 1989). However, it was not known whether this cholinergic potential is a property only of postmitotic sympathetic neurons, or of their embryonic SA progenitor as well. To address this question, the expression of ChAT mRNA and the synthesis and accumulation of ACh were examined in MAH cells. MAH cells contain low levels of ChAT mRNA, approximately 4 molecules per cell (Vandenbergh et al. 1991). However, they synthesize and store relatively high levels of ACh, almost a hundred times more than they do catecholamines (Table 1). Thus, MAH cells appear simultaneously noradrenergic and cholinergic. This mixed phenotype is not an artifact of the immortalized state of MAH cells, because primary SA progenitors also synthesize and accumulate ACh (Table 1), albeit at levels ten-fold lower than those measured in MAH cells (Vandenbergh et al. 1991).

These data suggest that the capacity to express a cholinergic phenotype is established early in the ontogeny of the SA lineage, prior to the decision between chromaffin and sympathetic neuronal fates. This early activation of cholinergic function may permit the progenitor to choose its transmitter phenotype in response to environmental signals as it migrates through the embryo. Consistent with this hypothesis, both MAH cells and primary SA progenitors are able to increase slightly (but reproducibly) their cholinergic function in

Table 1. Acetylcholine and catecholamine synthesis and accumulation[a]

Cell Type	CA (fmol/cell) × 10^4	ACh (fmol/cell) × 10^2	ACh/CA[b]	Fold Change[c]
MAH A3				
Dex	4.0 ± 0.5	3.8 ± 0.2	120 ± 10	1
Dex + L	4.5 ± 0.6	6.3 ± 0.4	170 ± 16	1.4[d]
MAHB2				
Dex	16 ± 1.0	2.8 ± 0.2	19 ± 2	1
Dex + LIF	16 ± 1.1	6.1 ± 0.2	39 ± 2	2.1[d]
E14.5 Adrenal cells				
Immediate labeling	380 ± 40	0.41 ± 0.05	0.11 ± 0.02	
No addition	580 ± 60	0.28 ± 0.03	0.052 ± 0.009	1
Dex	610 ± 30	0.28 ± 0.03	0.047 ± 0.006	0.9
LIF	470 ± 30	0.58 ± 0.05	0.12 ± 0.007	2.4[d]

[a] For MAH cells the data are expressed as the mean ± s.e.m. of 40 wells in 10 experiments (MAH A3 in Dex), 16 wells in 4 experiments (MAH A3 in Dex + LIF and MAH B2 in the transmitter content levels in MAH cells cultured without dex were highly variable, possibly reflecting the cell death that occurs under this condition. E 14.5 adrenal cells were analyzed after FACS sorting for HNK-1 staining and plating for 30 min. (Immediate labeling), or after 24 hr. in the described conditions. Data are expressed as the mean ± s.e.m. of 10 wells in 3 experiments (Immediate labeling), 6 wells in 2 experiments (No addition, Dex, and LIF). All incubations were for 4 hr. (From Vandenbergh et al. 1991)

[b] Ratios were calculated for each well in each experiment and the mean of these ratios is presented

[c] with respect to Dex

[d] Significant by Student's t-test to $p < 0.02$

response to CDF/LIF, a know cholinergic inducer (Table 1; Yamamori et al. 1989). This implies an early expression of receptors for CDF/LIF. MAH cells have also been shown to express receptors for ciliary neuronotrophic factor (CNTF; Ip et al, 1992), another factor known to induce cholinergic properties in SA lineage cells (Ernsberger et al. 1989; Rao and Landis 1990). Since SA progenitors generate chromaffin cells and sympathetic neurons which express little or no ACh synthesis, mechanisms must exist to down-regulate cholinergic properties during differentiation to these two cell types. The nature of these mechanisms is presently unknown.

To determine whether other neurotransmitter phenotypes might be available to the SA progenitor, the expression of two glutamic acid decarboxylase (GAD) mRNAs and tryptophan hydroxylase (TpH) mRNA was examined in MAH cells using a sensitive RNase protection assay (Vandenbergh et al. 1991). Neither GAD1 nor GAD2 mRNAs were detectably expressed in MAH cells. However, low levels of TpH transcripts (on the order of a few copies per cell) could be detected, using a PCR assay (Fig. 5). The lack of TpH mRNA expression in RAT1 fibroblasts (Fig. 5, lanes 1 and 2) indicates that expression

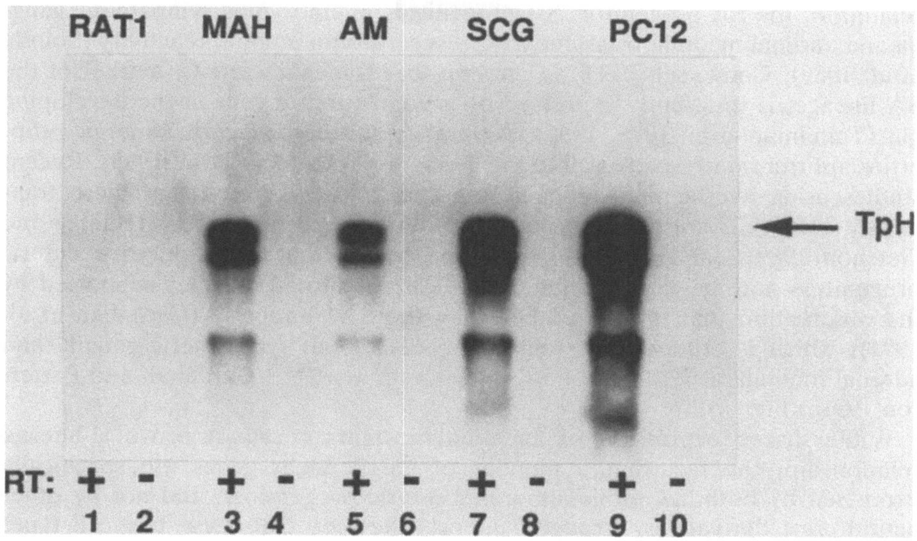

Fig. 5. Detection of tryptophan hydroxylase (TpH) mRNA by PCR amplification in cell lines and tissues from the sympathoadrenal lineage. Illustrated is a Southern blot of cDNA PCR-amplified from the tissues or cell types indicated, using primers specific for the TpH gene. The blot was hybridized with a TpH cDNA fragment that contains the amplified region as a control for the specificity of the amplification. As a further control, some RNA samples were subjected to cDNA synthesis in the absence (–) of reverse transcriptase (RT) and carried throug the procedure; as expected these controls lack a hybridizing band, indicating that the signals detected in odd-numbered lanes derive from reverse-transcribed mRNA and not from contaminating genomic DNA. cDNA from RAT1 fibroblast cells is negative; in a parallel control experiment this sample was amplified and hybridized with β_2 microglobulin sequences and yielded a signal of comparable intensity to that obtained from PC12 cells, indicating that equal amounts of template cDNA were present in these reactions (not shown). (After Vandenbergh et al. 1991).

of this gene, no matter how low, is lineage-specific. Moreover, the absence of detectable GAD mRNA expression suggests that not every neurotransmitter mRNA is expressed by the SA progenitor. The expression of TpH mRNA by MAH cells suggests that SA progenitors have the capacity to generate differentiated derivatives that synthesize and release serotonin. Consistent with this idea, sympathetic neurons have been shown to be capable of synthesizing serotonin in response to conditioned medium factors (Sah and Matsumoto 1987).

What is the potential biological significance of the capacity to express a serotonergic phenotype in SA progenitors? Serotonin is one of the neurotransmitter used by enteric neurons, which also derive from the neural crest (LeDouarin, 1982). The lineage of enteric neurons is not yet well-established. However, several converging lines of evidence strongly suggest that enteric neurons may derive from a progenitor that is identical, or at least closely related to, the SA progenitor. SA progenitors in embryonic sympathetic ganglia and adrenal medulla transiently express serotonin immunoreactivity (Soinila et al. 1989). Conversely, TH, an enzyme that is a characteristic marker of the SA lineage, is transiently expressed by a population of cells in the developing gut (Teitelman et al. 1978, 1981). A number of other catecholaminergic properties are transiently expressed in the gut as well (Jonakait et al. 1985). Recent studies using overlapping antigenic markers have suggested that these transiently TH^+ cells are likely to be progenitors of enteric neurons (Baetge and Gershon 1989; Baetge et al. 1990). The notion that these putative enteric progenitors may be closely related to SA progenitors is further supported by the observation that they are labeled by the SA1 antibody (Carnahan et al. 1991), which is otherwise restricted to cells of the sympathetic ganglia and adrenal medulla at E12.5–E14.5 (Anderson et al. 1991; Carnahan and Patterson 1991a; Fig. 6).

While shared expression of an antigenic marker cannot prove a lineage relationship, the fact that a number of independent genes are specifically expressed by both SA progenitors and enteric progenitors, and not by other neural crest derivatives, strongly supports the idea that these two cell types share a similar developmental program. In muscle, the developmental program

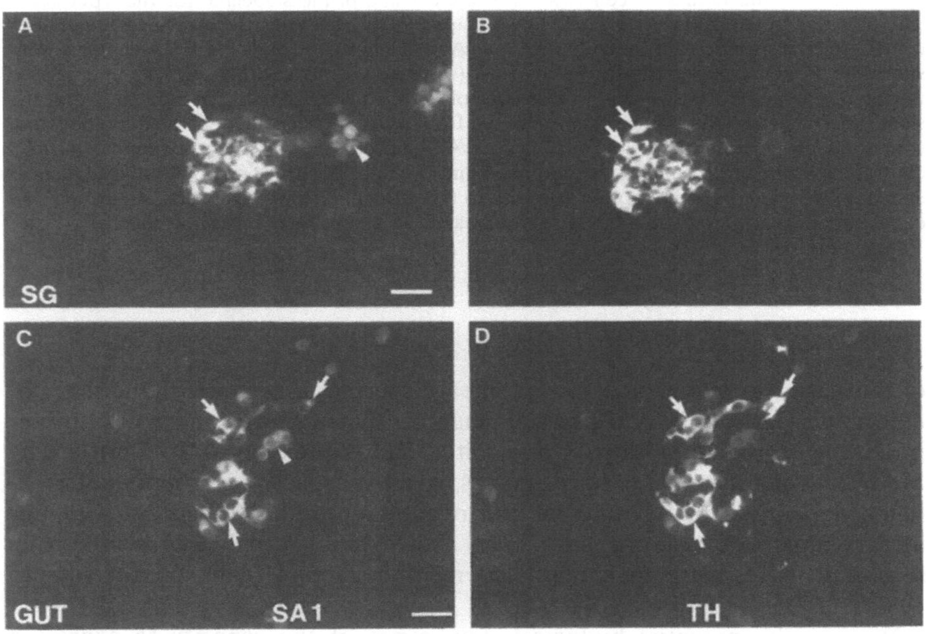

Fig. 6 A–D. Neuronal progenitors in sympathetic ganglia and the foregut express sympathoadrenal lineage-specific markers. Transverse sections through an E12.5 rat embryo are shown, double-labeled with monoclonal antibody SA1 (*A, C*) and polyclonal anti-TH (*B, D*). Individual cells in both the sympathetic ganglion (SG; A, B) and the gut (GUT; C, D) can be seen to co-express both markers (arrows). Arrowheads (A, C) indicate autofluorescent blood cells. Magnification bar 31 μM.

that leads to the coordinated expression of a battery of myocyte-specific genes is under the control of a family of homologous transcriptional regulatory molecules that includes MyoD (Weintraub et al. 1991). Recently, it has been shown that cells in the sympathetic ganglia primordia and the foregut both express MASH1 (Lo et al. 1991), a mammalian homologue of the *Drosophila* neurogenic determination genes *achaete-scute* (Johnson et al. 1990; Fig. 7). MASH1 and the *achaete-scute* genes belong to the same family of helix-loop-helix transcriptional regulators as MyoD (Johnson et al. 1990; Murre et al. 1989). If, as suggested, MASH 1 indeed functions as a vertebrate neurogenic determination gene (Lo et al. 1991), then the expression of this lineage-specific putative regulatory molecule in both sympathetic and enteric progenitors would strongly imply that these cells are similarly determined and lineally related. If such a lineage relationship were established, it would indicate that the SA progenitor is not simply a progenitor of chromaffin cells and sympathet-

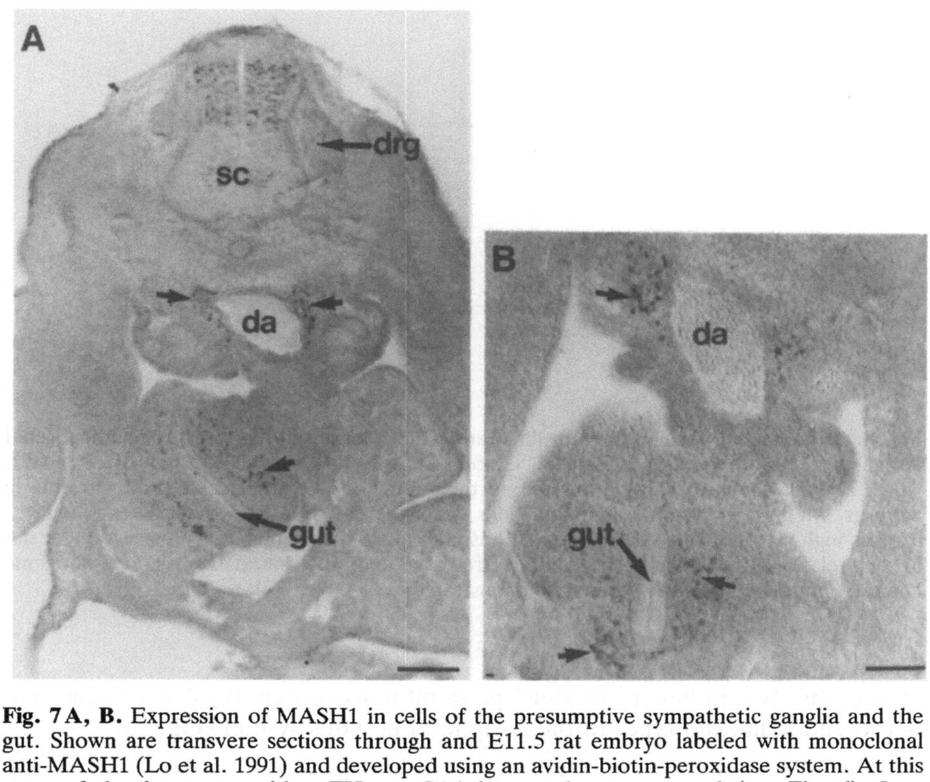

Fig. 7 A, B. Expression of MASH1 in cells of the presumptive sympathetic ganglia and the gut. Shown are transvere sections through and E11.5 rat embryo labeled with monoclonal anti-MASH1 (Lo et al. 1991) and developed using an avidin-biotin-peroxidase system. At this stage of development, neither TH nor SA1 has yet been expressed (see Fig. 6). Low magnification view (*A*) indicates MASH1-positive nuclei (arrows) aggregating adjacent to the dorsal aorta (da), the location of the sympathetic ganglion primordia, and scattered throughout the gut. MASH1-positive nuclei are also detected in the dorsal region of the spinal cord (sc). Note that the sensory neurons in the dorsal root ganglia (drg) are unlabeled by the antibody. Higher magnification view (*B*) illustrates more clearly the MASH1-positive nuclei (arrows) near the dorsal aorta and in the gut. Magnification bars: A, 128 μM; 64 μM.

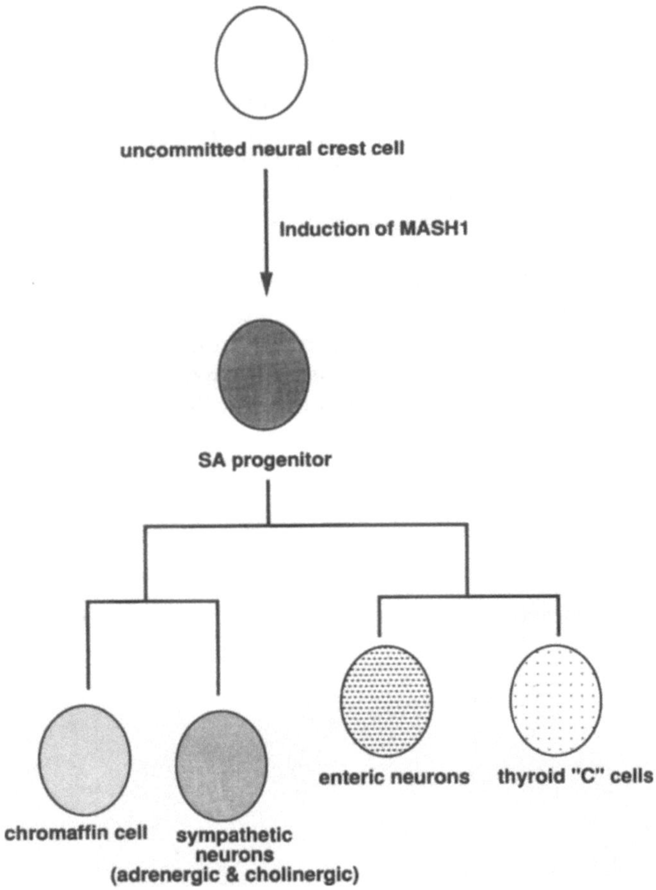

Fig. 8. Hypothetical lineage diagram illustrating the relationship between sympathoadrenal derivates and enteric neurons. Thyroid "C" cells are the endocrine cells of the thyroid gland, which are analagous to chromaffin cells in their function and their relationship to enteric neurons. Medullary thyroid carcinoma lines derived from these cells have been shown to express high levels of MASH1 mRNA (Johnson et al. 1990). It must be emphasized that the lineage relationships shown in this diagram are based upon the shared expression of antigenic markers, and have not been formally demonstrated.

ic neurons, but rather a "master" progenitor for a variety of endocrine and neuro-endocrine cell types in the peripheral nervous system (Fig. 8).

Implication and Limitations

The studies summarized in this article have focused on understanding the basic biology of one of the currently best-characterized neuronal progenitor cells in the vertebrate nervous system. This cell also happens to derive from a lineage from which cells have already been used in experimental therapies in humans. These studies have led to the conclusion that the SA progenitor is endowed with a rather broad repertoire of developmental potentials, and that its final choice of cell fate is exquisitely sensitive to the local environment. These conclusions may have more general applications in two different respects. First, the kinds of mechanisms that control sympathetic neuronal differentiation, such as growth factors relays, may operate on neuronal progenitors in the CNS as well (Cattaneo and McKay 1990). Second, the plasticity and broad developmental capacities of the SA progenitor could potentially make this cell useful for heterologous cell replacement therapies in regions of the nervous system developmentally unrelated to the neural crest, and attempts to test this idea are currently underway (Mahanthappa et al. 1990). However, it should be emphasized that our understanding of the SA progenitor is still relatively primitive. Moreover, because this cell generates both neurons and endocrine cells, the plasticity it exhibits may not be characteristics of other progenitors that give rise strictly to neurons (Rohrer and Thoenen 1987). Studies of progenitor cells in other neurogenic lineages will therefore be necessary to complement and expand the understanding of vertebrate neurogenesis derived from the study of this relatively simple model system.

Acknowledgments. The author would like to acknowledge the contributions of the individuals whose work is mentioned in this article, including S. J. Birren, D. L. Stemple, J. E. Johnson, L.-C. Lo, A. Michelsohn, D. J. Vandenbergh, A. S. Augsburger, N. K. Mahanthappa, J. F. Carnahan, and P. H. Patterson. In addition, he would like to thank Paul Patterson for many suggestions and intellectual contributions to the ideas discussed in this review. This work was supported by a Sloan Foundation Faculty Fellowship in Neurosciences, the PEW Foundation and the NIH.

References

Allen GS, Burns RS, Tulipan NB, Parker, RA (1989) Adrenal medullary transplantation to the caudate nucleus in Parkinson's disease. Arch Neurol 46:487–491

Anderson DJ (1988) Cell fate and gene expression in the developing neural crest. In: Gorio A (ed) Neural development and regeneration. NATO ASI Series, Vol. 22. Springer-Verlag, Berlin Heidelberg, pp 188–198

Anderson DJ (1989) The neural crest cell lineage problem: Neuropoiesis? Neuron 3:1–12

Anderson DJ, Axel R (1986) A bipotential neuroendocrine precursor whose choice of cell fate is determined by NGF and glucocorticoids. Cell 47:1079–1090

Anderson DJ, Michelsohn A (1989) Role of glucocorticoids in the chromaffin-neuron developmental decision. Int J Dev Neurosci 12:83–94

Anderson DJ, Carnahan J, Michelsohn A, Patterson PH (1991) Antibody markers identify a common progenitor to sympathetic neurons and chromaffin cells in vivo, and reveal the timing of committment to neuronal differentiation in the sympathoadrenal lineage. J Neurosci, 11:3507–3519

Baetge, G, Gershon MD (1989) Transient catecholaminergic (TC) cells in the vagus nerves and bowel of fetal mice: relationship to the development of enteric neurons. Dev Biol 132:189–211

Baetge G, Pintar JE, Gershon MD (1990) Transiently catecholaminergic (TC) cells in the bowel of the fetal rat: precursors of noncatecholaminergic enteric neurons. Dev Biol 141:353–380

Baroffio A, Dupin E, LeDouarin NM (1988) Clone-forming ability and differentiation potential of migratory neural crest cells. Proc Natl Acad Sci USA 85:5325–5329

Beato M (1989) Gene regulation by steroid hormones. Cell 56:335–344

Birren SJ, Anderson DJ (1990) A v-myc-immortalized sympathoadrenal progenitor cell line in which neuronal differentiation is initiated by FGF but not NGF. Neuron 4:189–201

Birren SJ, Verdi, J and Anderson, DJ (1992). Membrane depolarization induces p140^{trk} and NGF-responsiveness, but not P75,^{LN6FR} in MAH cells. Science 257:395–397

Björklund A (1991) Neural transplantation – an experimental tool with clinical possibilities. Trends Neurosci 14:319–322

Bohn MC, Goldstein M, Black I (1981) Role of glucocorticoids in expression of the adrenergic phenotype in rat embryonic adrenal gland. Dev Biol 82:1–10

Bronner-Fraser M, Fraser S (1989) Developmental potential of avian trunk neural crest cells in situ. Neuron 3:755–766

Brundin P, Nilsson OG, Strecker RE, Lindvall O, Astedt B, Björklund A (1986) Behavioural effects of human fetal dopamine neurons grafted in a rat model of Parkinson's disease. Exp Brain Res 65:235–240

Carnahan J, Patterson PH (1991a) The generation of monoclonal antibodies that bind preferentally to adrenal chromaffin cells and the cells of ambyonic sympathetic ganglia. J Neurosci 11:3493–3506

Carnahan J, Patterson PH (1991 b) Isolation of the progenitor cells of the lineage from embyonic sympathetic ganglia with the SA monoclonal antibodies. J Neurosci 11:3520–3530

Carnahan JF, Anderson DJ, Patterson PH (1991). Evidence that enteric neurons may derive from the sympathoadrenal lineage. Dev Biol 148:552–561

Cattaneo E, McKay RDG (1990) Proliferation and differentiation of neuronal stem cells regulated by nerve growth factor. Nature 347:762–765

Cattaneo E, McKay R (1991) Identifying and manipulating neuronal stem cells. Trends Neurosci 14:338–340

Claude P, Parade IM, Gordon KA, D'Amore PA, Wagner JA (1988) Acidic fibroblast growth factor stimulates adrenal chromaffin cells to proliferate and to extend neurites, but is not a long-term survival factor. Neuron 1:783–790

Doupe AJ, Patterson PH, Landis SC (1985) Environmental influences in the development of neural crest derivatives: glucocorticoids, growth factors and chromaffin cell plasticity. J Neurosci 5:2119–2142

Ernsberger V, Sendtner M, Rohrer H (1989) Proliferation and differentiation of embryonic chick sympathetic neurons: effects of ciliary neurotrophic factor. Neuron 2:1275–1284

Freed WJ, Poltorak M, Becker JB (1990) Intracerebral adrenal medulla grafts: A review. Exp Neurol 110:139–166

Fukada K (1985) Purification and partial characterization of a cholinergic neuronal differentiation factor. Prog Natl Acad. Sci USA 82:1895–1899

Gage FH, Buzsáki G (1989) CNS grafting: potential mechanisms of action. In: Neural regeneration and transplantation. Alan R Liss Inc, New York, pp 211–226

Gage FH, Wolff JA, Rosenberg MB, Xu L, Yee J-K, Shults C, Friedmann T (1987) Grafting genetically modified cells to the brain: possibilities for the future. Neuroscience 23:795–807

Gage FH, Kawaja MD, Fisher LJ (1991) Genetically modified cells: applications for intracerebral grafting. Trends Neurosci 14:328–333

Hempstead BL, Martin-Zanca D, Kaplan DR, Parada LF, Chao MV (1991) High-affinity NGF binding requires coexpression of the trk protooncogene and the low-affinity NGF receptor. Nature 350:678–683

Heuer JG, von Bartheld CS, Kinoshita Y, Evers PC, Bothwell M (1990) Alternating phases of FGF receptor expression and NGF receptor expression in the developing chicken nervous system. Neuron 5:283–296

Ip NY, Nye SH, Boulton TG et al. (1992). CNTF and LIF act on neuronal cells via shared signalling pathways that involve the IL-6 signal transducing component gp130. Cell 69:1121–1132

Johnson JE, Birren SJ, Anderson DJ (1990) Two rat homologues of Drosophila achaete-scute specifically expressed in neuronal precursors. Nature 346:858–861

Jonakait GM, Markey KA, Goldstein M, Dreyfus CF, Black IB (1985) Selective expression of high-affinity uptake of catecholamines by transiently catecholaminergic cells of the rat embryo: studies in vivo and in vitro. Dev Biol 108:6–17

Kaplan DR, Hempstead BL, Martin-Zanca D, Chao MV, Parada LF (1991 a) The trk proto-oncogene product: a signal transducing receptor for nerve growth factor. Science 252, 554–558

Kaplan DR, Martin-Zanca D, Parada LF (1991 b) Tyrosine phosphorylation and tyrosine kinase activity of the trk proto-oncogene product induced by NGF. Nature 350:158–160

Klein R, Jing S, Nanduri V, O'Rourke E, Barbacid M (1991) The trk protooncogene encodes a receptor for nerve growth factor. Cell 65:189–197

LeDouarin NM (1982) The neural crest. Cambridge University Press, Cambridge, UK

Lindvall O, Brundin P, Widner H, Rehncrona S, Gustavii B, Frackowiak R, Leenders KL, Sawle G, Rothwell JC, Mardsen CD, Björklund A (1990) Grafts of fetal dopamine neurons survive and improve motor function in Parkinson's disease. Science 247:574–577

Lo L, Johnson JE, Wuenschell CW, Saito T, Anderson DJ (1991) Mammalian achaete-scute homolog 1 is transiently expressed by spatially-restricted subsets of early neuroepithelial and neural crest cells. Genes Dev 5:1524–1537

Manhanthappa NK, Gage FH, Patterson PH (1990) Adrenal chromaffin cells as multipotential neurons for autografts. Progr Brain Res 82:33–39

Martin-Zanca D, Hughes SH, Barbacid M (1986) A human oncogene formed by the fusion of truncated tropomyosin and protein tyrosine kinase sequences. Nature 319:743–748

Mori N, Birren SJ, Stein R, Stemple D, Vandenbergh DJ, Wuenschell CW, Anderson DJ (1990) Contributions of cell-extrinsic and cell-intrinsic factors to the differentiation of a neural crest-derived neuroendocrine progenitor cell. Cold Spring Harbor Symp. Quant. Biol. LV, 255–264

Michelsohn A, Anderson DJ (1992) Changes in competence determine the timing of two sequential glucocorticoid effects on sympathoadrenal progenitors Neuron 8:589–604

Mori N, Birren SJ, Stein R, Stemple D, Vandenbergh DJ, Wuenschell CW, Anderson DJ (1990) Contributions of cell-extrinsic and cell-intrinsic factors to the differentiation of a neural crest-derived neuroendocrine progenitor cell. Cold Spring Harbor Symp Quant Biol LV 255–261

Murre C, McCaw PS, Baltimore D (1989) A new DNA binding and dimerization motif in immunoglobin enhancer binding, daughterless, MyoD and myc proteins. Cell 56:777–783

Patterson PH (1990) Control of cell fate in a vertebrate neurogenic lineage. Cell 62:1035–1038

Patterson PH, Chun LLY (1977) The induction of acetylcholine synthesis in primary cultures of dissociated rat sympathetic neurons. I. Effects of conditioned medium. Dev Biol 56:263–280

Rao MS, Landis SC (1990) Characterization of a target-derived neuronal cholinergic differentiation factor. Neuron 5:899–910

Rohrer H, Thoenen H (1987) Relationship between differentiation and terminal mitosis: chick sensory and ciliary neurons differentiate after terminal mitosis of precursor cells, whereas sympathetic neurons continue to divide after differentiation. J Neurosci 7:3739–3748

Sah DWY, Matsumoto SG (1987) Evidence for serotonin synthesis, uptake and release in dissociated rat sympathetic neurons in culture. J Neurosci 7:391–399

Seidl K, Unsicker K (1989) The determination of the adrenal medullary cell fate during embryogenesis. Devl Biol 136:481–490

Soinila S, Ahonen M, Lahtinen T, Häppölä O (1989) Developmental changes in 5-hydroxy-tryptamine immunoreactivity of sympathetic cells. Int J Dev Neurosci 7:553–563

Stemple DL, Mahanthappa NK, Anderson DJ (1988) Basic FGF induces neuronal differentiation, cell division, and NGF dependence in chromaffin cells: a sequence of events in sympathetic development. Neuron 1:517–525

Teitelman G, Joh TH, Reis DJ (1978) Transient expression of a noradrenergic phenotype in cells of the rat embryonic gut. Brain Res 158:229–234

Teitelman G, Gershon MD, Rothman TP, Joh TH, Reis DJ (1981) Proliferation and distribution of cells that transiently express a catecholaminregic phenotype during development in mice and rats. Dev Biol 86:348–355

Unsicker K, Drisch B, Otten J, Thoenen H (1978) Nerve growth factor-induced fiber outgrowth from isolated rat adrenal chromaffin cells: impairment by glucocorticoids. Proc Natl Acad Sci USA 75:3498–3502

Vandenbergh DJ, Mori N, Anderson DJ (1991) Co-expression of multiple neurotransmitter enzyme genes in normal and immortalized sympathoadrenal progenitor cells. Dev Biol 148:10–22

Weintraub H, Davis R, Tapscott S, Thayer M, Krause M, Benezra R, Blackwell TK, Turner D, Rupp R, Hollenberg S, Zhuang Y, Lassar A (1991) The myoD gene family: nodal point during specification of the muscle cell lineage. Science 251:761–766

Yamamori Y, Fukada K, Aebersold R, Korsching S, Fann M-J, Patterson PH (1989) The cholinergic neuronal differentiation factor from heart cells is identical to leukemia inhibitory factor. Science 246:1412–1416

Lineage, Migration, and Phenotype in Avian Optic Tectum: Analysis with Recombinant Retroviral Vectors

J. R. Sanes, S. Galileo, I. R. Ghattas, G. E. Gray, and *J. E. Majors*

Summary

We have used retrovirus mediated gene transfer in two ways to analyze developmental processes in chicken optic tectum. First, viruses encoding the marker gene, *lacZ,* were employed as tracers to analyze the lineage and migration of tectal cells. These studies demonstrated that a multipotential progenitor gives rise to neurons, astrocytes, and radial glial cells, and revealed a striking relationship between the migratory path that a cell follows and the phenotype it acquires. Second, with the aim of assessing roles that specific genes play, we have begun to use retroviral vectors to coexpress lacZ and a potentially bioactive gene, or its antisense copy, in clones of cells. In initial studies, overexpression of an oncogene (v-src) and antisense suppression of an adhesion molecule (integrin β1) have perturbed the behavior of lacZ-positive cells in ways that provide insights into the functions of these genes.

Introduction

Most methods currently used to express exogenous genes within the nervous system aim to affect large numbers of cells. These include the generation of transgenic or chimeric animals, the implantation of genetically modified cells, and infection with vaccinia or herpes virus vectors. However, there are some applications for which it is important to target only small numbers of cells. The most obvious of these is lineage analysis: insertion of a stable, preferably heritable, marker into individual progenitors allows their clonal progeny to be identified so that genealogical relationships can then be determined. In addition, if one could alter the expression of potentially bioactive genes in small numbers of cells without otherwise perturbing the environment, cell-autonomous and/or direct effects of the transgene could be assessed.

With these aims in mind we have explored the use of recombinant retroviral vectors as agents for gene transfer *in vivo*. When a retrovirus infects a cell, its genome is reverse transcribed and inserted into the host genome; thereafter, the inserted genes are replicated, transcribed and translated along with host genes (Weiss et al. 1982). Thus, at least in principle, retrovirally transferred genes will be stably expressed in all of the progeny of the initially infected cells.

By constructing replication-defective vectors and using small numbers of virions, it is possible to limit the transferred genes to a minuscule fraction of the cells within a target area. Incorporation of a suitable "marker" gene into the recombinant vector allows the progeny of infected cells to be identified so that their locations can be mapped and their phenotypes assessed. Here, we summarize recently completed analysis of neural cell lineage and recently initiated attempts to perturb neural phenotype, both of which rely heavily on the use of such recombinant retroviral vectors.

Lineage analysis

The first set of vectors we constructed were designed for lineage analysis. Genes encoding viral proteins *(gag, pol,* and *env)* were excised from the retroviral genome and replaced by the *E.coli* β-galactosidase *(lacZ)* gene. Removal of essential viral genes rendered the retrovirus replication-defective. Because lacZ (we call the gene *lacZ* and its product lacZ) can be detected histochemically, individual descendants of cells infected with these vectors can be labeled in whole mounts or tissue sections. In practice, a plasmid encoding the recombinant retroviral genome is transfected into a "helper" cell which stably expreses retroviral structural genes. Viral proteins and cellular lipids package the recombinant genomes, which are then secreted into the culture medium as active virions (Fig. 1a). First generation helper cells provided viral genes from a retroviral genome that could not itself be packaged because its "packaging signal" was mutated. More recently designed helper cells supply viral proteins from more profoundly disrupted viral genomes to minimize the incidents of recombinational events that could lead to the regeneration of replication competent virions (Miller and Rosman 1989). In either case, the virus is collected by centrifugation and an aliquot containing a small number of virions is injected into an embryo at an appropirate stage and place. After an appropriate interval, the region to be analyzed is fixed and stained for lacZ. In some cases, tissue is first sectioned and then stained; in other cases, whole mounts are stained and clones of lacZ-positive cells are subsequently isolated and sectioned.

In most of the tissues that we have analyzed we found coherent clusters of lacZ-positive cells surrounded by vast expanses of lacZ-negative tissue. We base our clonal analysis on the assumption that each cluster represents the progeny of a single infected cell. The validity of this assumption is supported by three arguments. First, we routinely inject a small number of virions and restrict our analysis to cases in which only a few clusters of lacZ-positive cells occur. Given the injection protocol and the labeling pattern observed, it seems unlikely that separate virions would regularly infect neighboring cells and thus produce polyclonal clusters (Sanes et al., 1986). Second, we perform dilution experiments in which a small fraction of the normal viral dose is applied. If each cluster is a clone, one would expect that a 10-fold diluted solution would give rise to 10% as many clones as the normal dose, but that each cluster would contain on average the same number of cells. If, however, many clusters were

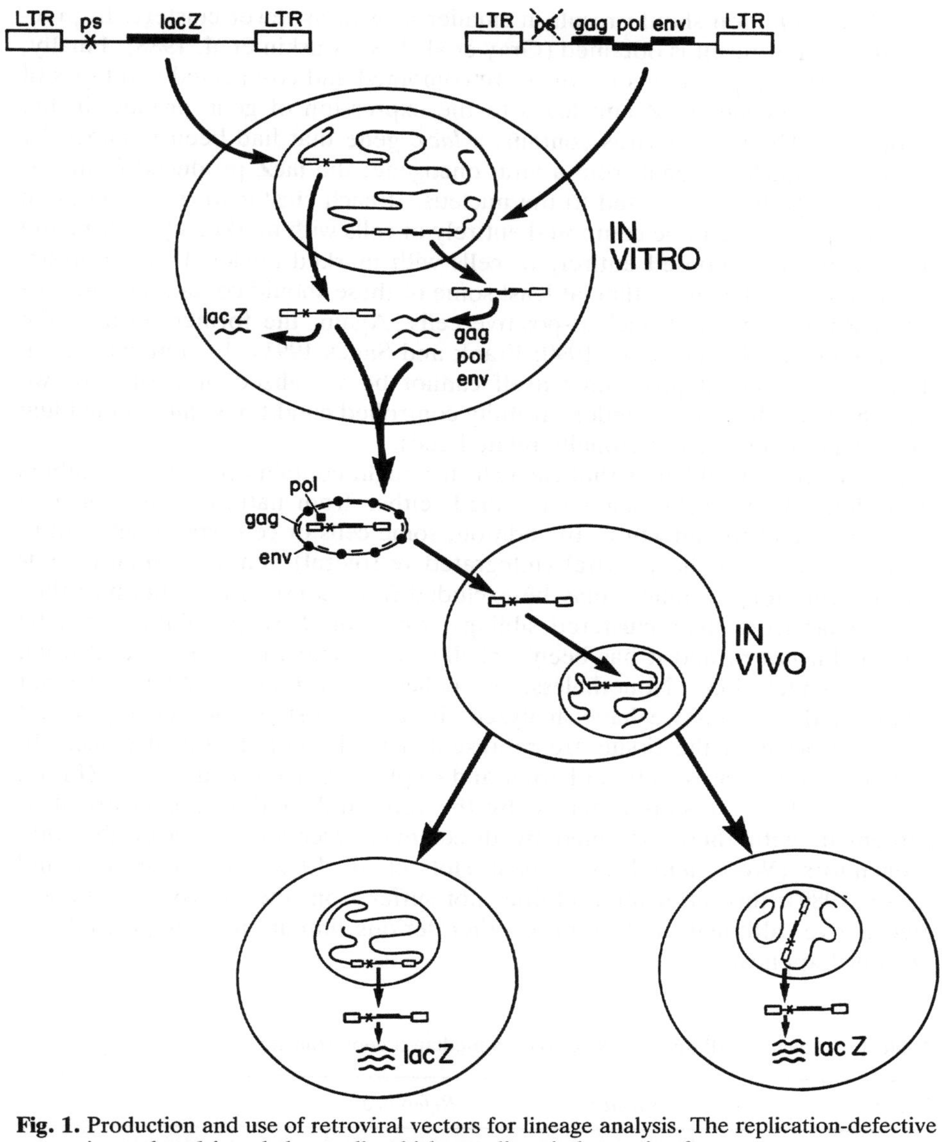

Fig. 1. Production and use of retroviral vectors for lineage analysis. The replication-defective vector is produced in a helper cell, which supplies viral proteins from a separate genome; both of these vectors are introduced into the producer cell by transfection. Virions are then harvested from the medium and injected into embryos. There, the retroviral genome is stably incorporated into a chromosome of a progenitor cell and passed to that cell's progeny, wherein it directs expression of the marker, lacZ. (Adapted from Sanes, 1989a).

polyclonal, dilution should result in smaller as well as fewer clusters. In practice the former result is obtained (Gray et al. 1988; Luskin et al. 1988). Finally, we performed experiments in which we combined and co-injected two types of virus. One encodes *lacZ* and leads to the expression of gene product in the cytoplasm. The second virus contains a *lacZ* gene that had been modified by fusion to a nuclear signal from a viral oncogene; the lacZ produced from this vector is rapidly translocated to the nucleus. If each cluster were a clone, one would expect some to be composed entirely of cells with marked cytoplasm and the rest to be composed entirely of cells with marked nuclei. If, in contrast, clusters were polyclones, then at least some of these should contain mixtures of cytoplasm-positive and nucleus-positive cells. Again, the former result is the one obtained (Galileo et al. 1990; Frank and Sanes 1991). In summary, even though the infected progenitor itself cannot be visualized or identified, we nonetheless believe that, under carefully controlled conditions, most individual clusters are composed of clonally related cells.

The converse problem is that each cluster might contain only some members of a clone. Other cells might have died, either from natural causes or as a consequence of the infection. In addition, some cells or cell types might fail to express lacZ from the proviral (integrated retroviral) genome. Finally, cells that migrate long distances may be excluded from consideration because they end up far from their clustered siblings. None of these possibilities can be excluded *a priori,* and it has been a challenge to bear them all in mind when drawing conclusions. Nonetheless, we believe that these problems are not major in the tissues we have analyzed, because most of the cell types that normally occur in the tissue are represented in the population of clones. In addition, in two cases, retina (Turner and Cepko 1987) and neural crest (Frank and Sanes 1991), results obtained by the retroviral method are in excellent agreement with those obtained by direct microinjection of tracer dye into progenitors (Wetts and Fraser 1988; Holt et al. 1988; Bronner-Fraser and Fraser 1989); this latter method does not suffer from the drawbacks enumerated above, although it does have other serious limitations that preclude its universal application.

Table 1. Studies of cell lineage by retrovirus-mediated gene transfer

Tissue	*Species*	*Reference*
Skin	mouse	Sanes et al. 1986
Yolk sac	mouse	Sanes et al. 1986
Retina	mouse and rat	Turner and Cepko 1987; Turner et al. 1990
Cerebral cortex	mouse and rat	Luskin et al. 1988; Walsh and Cepko 1988; Price and Thurlow 1988
Peripheral nerve	rat	Bunge et al. 1989
Optic tectum	chick	Gray et al. 1988; Galileo et al. 1990; Gray and Sanes 1991, 1992
Forebrain	chick	Gray et al. 1990
Dorsal root ganglion	chick	Frank and Sanes 1991
Spinal cord	chick	Leber et al. 1990

Over the past several years, the retroviral method has been used to analyze cell lineage in a number of neural and nonneural tissues, both in rodents and in birds (Table 1). To illustrate the types of results that this method can generate, and to provide a basis for a evaluating our more recent studies of gene transfer, we next summarize results from the tissue on which we have focussed most recently, the chicken optic tectum.

The optic tectum

The optic tectum is the major center for processing visual information in birds. As such, and because it is large and accessible at early embryonic stages, its structure and development have been studied in some detail (see Gray et al. 1988; Gray and Sanes 1991 for references). Moreover, the tectum displays several intriguing organizational features to which patterns of cell lineage and migration might contribute. First, input from the retina forms an orderly topographic (retinotopic) map onto the tectal surface; indeed, this map has been the subject of the majority of studies on tectal development to date. Second, the tectum is divided into some 15 distinct laminae, which run parallel to each other and to the pial surface. Each lamina bears a distinct constellation of inputs, outputs, neuronal morphologies, and neurotransmitter phenotypes. Third, there is a radial component to the functional organization of the tectum, such that radially neighboring neurons in different laminae are likely to be functionally related and synaptically interconnected (Jassik-Gerschenfeld and Hardy 1984). Interestingly, the tectum shares all of these organization features with the mammalian cerebral cortex, and it seems likely that tectum and cortex use at least some similar developmental strategies.

To study lineage in the tectum, we injected retrovirus into the tectal vesicle on embryonic days (E) 2 and 3, which is just as the first neurons are born (LaVail and Cowan 1971). We then analyzed tecta late in embryogenesis (E15-19; hatching is at E20) after all neurons were born and the definitive laminae had formed. Our two main results were:

1. Individual clones frequently contained neurons of several types that resided in several laminae; interneurons and projection neurons regularly appeared in the same clones. Thus, individual progenitors can give rise to multiple neuronal subtypes (Gray et al. 1988).
2. By using methods of immunohistochemical double-labeling, we identified astrocytes in many of the clones that contain neurons.

Individual clones frequently contained either or both of two immunohistochemically distinct astrocyte subtypes as well as neurons (Galileo et al. 1990). Together these results are most consistent with the idea that a single type of progenitor gives rise to a wide range of tectal cell types, including output neurons, interneurons, and two types of astrocytes (Fig. 2a). These results do not, in themselves, reveal whether tectal lineage resembles that in the retina (Fig. 2b, right), where even the terminal division can produce a neuron and a glial cell (Turner and Cepko 1987; Wetts and Frasier 1988; Holt et al. 1988), or whether the tectum is more similar to the mammalian cerebral cortex (Fig. 2b, left), in which neuronal and glial lineages diverge to give rise to discrete populations of neuronal and astrocytic progenitors (Luskin et al. 1988; Price and Thurlow 1988). However, studies in which recombinant retrovirus was

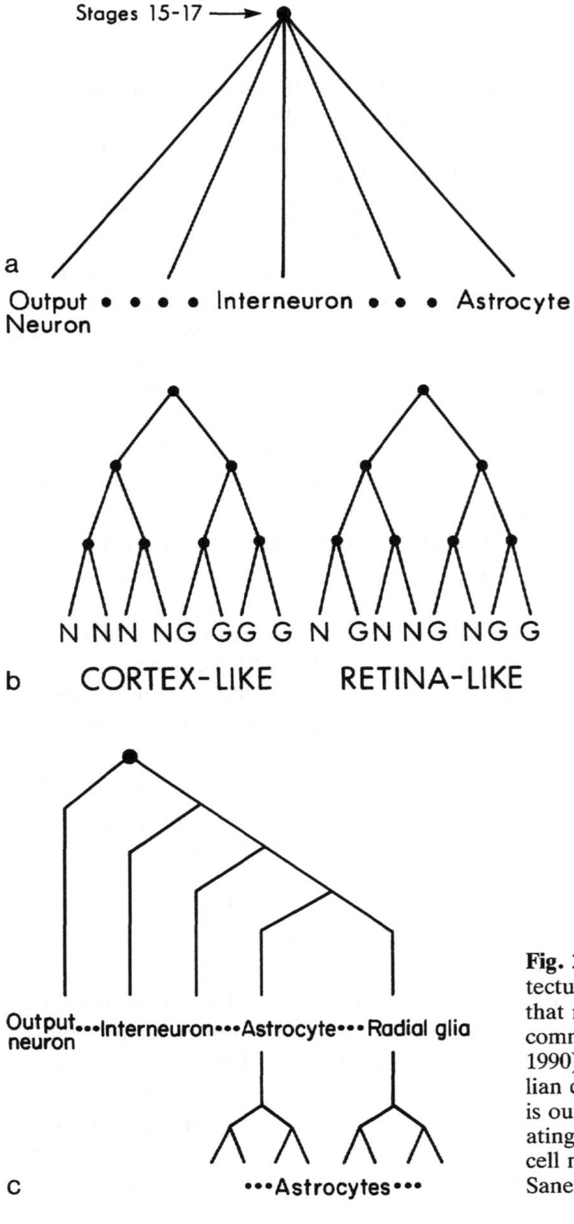

Fig. 2a–c. Cell lineage in chick optic tectum. a summarizes results showing that neurons and astrocytes arise from a common progenitor (Galileo et al. 1990). b presents results from mammalian cortex and retina, for comparison. **c** is our current lineage diagram, incorporating data on radial glia and on a stem cell mode of proliferation (see Gray and Sanes 1991, 1992).

injected into tectum at successively later times to generate successively smaller clones have not yet provided any evidence for a divergence of the lineages, and therefore favor the retinal model.

More recently, we have extended the lineage analysis to encompass another cell type, the radial glial cell (Gray and Sanes 1992; Fig. 2c) As will be discussed further below, the radial glial cell has been studied intensively in the mammalian cerebral cortex and shown to be an early born and transient cell type that guides the migration of neuroblasts from the ventricular zone to their final laminar destinations (Rakic 1988; Hatten 1990). In tectum, we used a combination of morphological and immunohistochemical features to identify radial glial cells. We found that clones containing radial glia frequently contained neurons and/or astrocytes, but usually no other radial glia. Thus, radial glia arise from a multipotential progenitor rather than from a committed radial glial precursor. Interestingly, production of radial glia continues at least until E8, after the peak of neuronal birth is over. Radial glia and neuronal lineages do not appear to diverge during this interval, and radial glia may be among the last cells that their progenitors produce. Finally, the population of labeled radial glia declines late in embryogenesis (between E15 and E19), concurrent with a sharp increase in the number of labeled astrocytes. This result, as well as the occurrence of lacZ-positive radial glia-astrocyte "transitional forms", suggests that some tectal radial glia transform into astrocytes after their migratory function is complete, as is thought to occur in mammalian cerebral cortex (Voigt 1989). In contrast to cortex, however, many tectal radial glia persist into maturity (Vanselow et al. 1989). To reconcile the observations that many radial glia are present early, that radial glia are among the last offspring of a multipotential stem cell, and that most clones contain only a single radial glial cell, we suggest that the stem cell is, or becomes, a radial glial cell (see also Fredricksen and McKay 1988 and Alvarez-Buylla et al. 1990).

In light of the evidence that a major tectal progenitor is multipotential, it seemed important to begin looking for extrinsic factors that might act on uncommitted cells to influence their phenotypic choices. In this context, it is logical to consider the environment that cells encounter as they migrate from their birthplaces to their laminar destinations. As is the case in most portions of the central nervous system, neurons and astrocytes all arise in a neuroepithelial layer called the ventricular zone. They then migrate apically to form and populate the laminae. In our initial studies, we found that most (but not all) lacZ-positive cells within each clone formed a compact radial array as they migrated and that most membres of most clones remained radially arranged even after migration was complete (Gray et al. 1988). These results showed that cells derived from a common progenitor followed common radial guides to their destinations, and suggested a possible relationship between genealogy and the radial functional organization that characterizes the mature tectum. To extend these studies, we injected large numbers of embryos with virus at E3 and then studies tecta at multiple, closely spaced stages. By mapping the postions of clonally related cells at each stage we assembled a composite picture of the migratory routes chosen by clonal relatives.

The results of this study are summarized in Figure 3 (Gray and Sanes 1991). Initially, stem cells divide rapidly and their progeny all stack radially in the ventricular zone. On E6, however, a few cells leave the radial array apically and begin to migrate tangentially just beneath the pial surface. This tangential migration is oriented in the mediolateral direction, and appears to represent the movement of neuroblasts along oriented fascicles of axons that themselves arise from some of the first-born tectal neurons. The number of tangential migrants derived from each clone is small, as is the time interval during which emigration occurs. After E7, the majority of cells proceed radially through an intermediate zone to form a tectal plate. The radial migration occurs along, and is presumably guided by, fascicles of radial glial cells, which interdigitate with axonal fascicles in the intermediate zone. Interestingly, clones of cells are frequently aligned along a glial fascicle that includes the apical process of a clonally related radial glial cell; thus, radial glia may guide the migration of their siblings. After the radial migration has begun to populate the tectal plate, a second group of cells leaves the radial array beneath the pial surface. Unlike the first tangential migration, however, the second is not obviously oriented along any single axis. One possibility is that this migration is stimulated but not oriented by the retinal axons, which invade the pial surface of the tectum at about this time. Then, during the last ten days of embryogenesis, all three groups of cells continue to migrate as the tectum expands, and all acquire distinctive phenotypes. Virtually all of the cells in the first tangential migration differentiate into multipolar ganglion cells, which are the main output neurons of the tectum and among the first to be born. Many, although not all, of the cells in the second tangential migration differentiate into subpial astrocytes. The cells that remain radially arrayed, which constitute the majority, differentiate into a variety of neuronal subtypes as well as into astrocytes.

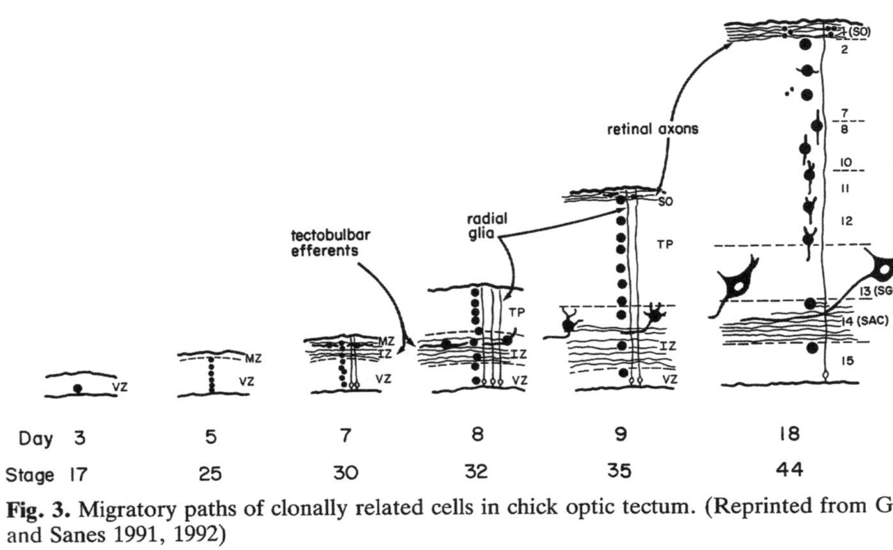

Fig. 3. Migratory paths of clonally related cells in chick optic tectum. (Reprinted from Gray and Sanes 1991, 1992)

Thus, our descriptive studies of tectal development revealed an unexpected relationship between migratory path and phenotypic choice. Cells descended from a single multipotential progenitor begin to migrate together but subgroups then diverge to form three discrete migratory streams which are guided by three sets of extrinsic cues and which adopt distinct sets of phenotypes. These results, in turn, raise two new questions. First, how are clonally related cells sorted onto their correct pathways? One possibility is that there are three sets of molecules that mediate movement of cells along the three streams; if so, it is important to know what these are and how their expression is regulated. Second, is there a causal relationship between a cell's destination and its phenotype? For example, might a particular neuroblast become a multipolar neuron because it has migrated tangentially or, alternatively, might a neuroblast migrate tangentially because it is already committed to become a multipolar cell? A general approach to both of these issues is to misexpress specific candidate genes within individual clones of migrating tectal cells. One could then ask whether particular genes are involved in the migratory process, and whether perturbatrion of migratory path affects phenotype. With this as a long-term goal, we have begun to design and test retroviral vectors capable of transferring *lacZ* plus a second gene (or its antisense copy) to tectal progenitors.

Retroviral vectors that coexpress two genes

Retroviruses have been successful in evolving strategies for coexpressing several of their own genes from a single genome. These include synthesis and subsequent processing of fusion proteins, ribosomal frame shifting, and regulated splicing to generate subgenomic messages (Weiss et al. 1985). In addition, numerous recombinant retroviral vectors have been constructed that contain not only a gene of interest (for example a marker gene or a candidate for gene therapy) but also a selectable marker (Friedmann 1989). We therefore expected that it would be straightforward to design vectors that would transfer both *lacZ* and a bioactive gene to tectal progenitors. Unexpectedly, when we constructed such vectors and tested them in culture, we found that most infected cells expressed one gene or the other and that only a minority expressed both. Possible explanations for this outcome are as follows:

First, the vectors we tested initially, like most of those being developed for gene therapy, use two separate promoters to direct expression of the two genes (see Fig. 4a). The upstream gene is expressed from the retroviral promoter in the long terminal repeat, and the downstream gene is expresed from an internal promoter. Unfortunately, such vectors are often compromised by a problem of competitive interference between promoters (Emmerman and Temin 1986). The extent of the interference is dependent on the strength of the promoter and the context in which it is placed. Although the molecular basis of the interference is still uncertain, many cases have been documented in which individual clones of infected cells express one gene or the other rather than both, even though they all harbor the same proviral genome.

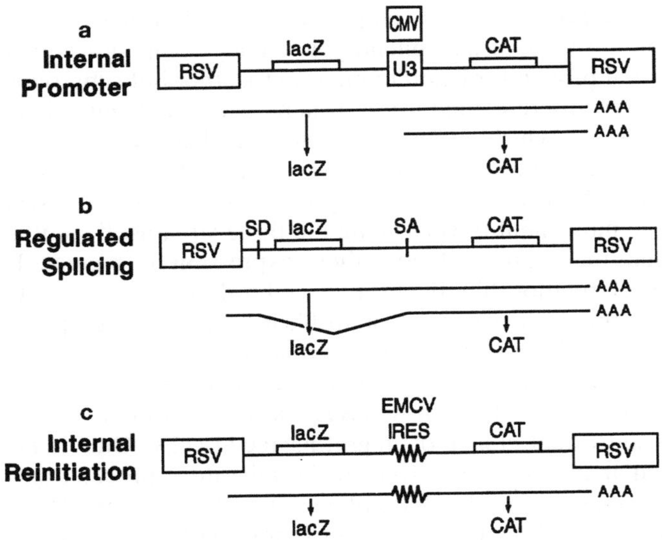

Fig. 4a–c. Alternative strategies for expressing two-genes (LacZ and CAT, in this example) from a single retroviral genome. See text for details.

An alternative strategy for coexpression of two genes involves the generation of separate mRNAs by regulated splicing of the single genomic transcript (see Fig. 4b). This strategy mimics that used by retroviruses to generate endogenous gene products such as *env* and *v-src*. With this approach, however, expression of one gene is always at the expense of the other. Moreover, regulation of alternative splicing is not well understood in any system and it is clear that splicing efficiency is highly context-dependent (Arrigo and Beemon 1988; Stoltzfus and Fogarty 1989). It is therefore currently infeasible to engineer vectors that produce spliced and unspliced RNAs in optimal (or even predictable) ratios.

In an attempt to circumvent these limitations, we have tested a third strategy for expressing multiple genes from a single proviral genome (Fig. 4c). Our approach was based on the recent demonstration that the 5' nontranslated region of several picornaviruses contained sequences that allow cap-independent initiation of translation from an internal ribosome entry site (IRES; Jackson et al. 1990). By inserting an IRES from the encephalomyocarditis virus between discrete coding regions within a retroviral genome, we constructed vectors that produce a single bicistronic transcript from which two gene products can be translated. In this way, transcriptional controls and RNA processing steps that differentially affect expression of multiple genes can, in principle, be avoided.

As an initial test of this strategy, we compared a series of vectors that encoded both *lacZ* and CAT (Ghattes et al. 1991; Fig. 4). *lacZ* was the upstream gene in all of the vectors and was expressed from the long terminal repeat. CAT was expressed in one of six ways: from a spliced subgenomic message, from an internal Rous sarcoma virus promoter, from an internal cytomegalovirus promoter, or from three variant IRES fragments. Virus particles genrated from each of the plasmids were used to infect cultured cells. Subsequently, the infected cells were assayed in three ways. First, we estimated the viral titers generated by each plasmid. Titers were consistently higher from IRES vectors than from the other vectors. This result is consistent with the possibilities that negative interactions between promoters inhibit the production of genomic RNA and that expression of spliced subgenomic mRNA is at the expense of full-length genomic RNA. Second, we estimated levels of CAT and lacZ expression by assaying both enzymes in extracts prepared from the infected cells. We found all vectors expressed appreciable levels of both genes, but that the ratio of CAT to lacZ was higher from IRES vectors than from the double promoter or regulated splicing vectors. Finally, to address the issue of coexpression at the cellular level, we used immunohistochemical double staining to ask how many of infected cells expressed both lacZ and CAT. As shown in Table 2, levels of coexpression were high (> 90%) with all of the IRES vectors, whereas the majority of cells infected by regulated splicing or double promoter vectors expressed either CAT or lacZ but not both. Thus, the IRES appears to provide the direct best available system for reliable coexpression of these two genes from a single proviral genome.

To test the general utility of the IRES approach, we constructed additional sets of vectors in which either the oncogene *v-src* or the selectable marker *neo*^{R} was placed downstream of the *lacZ* gene. In both cases, the IRES vectors provided an efficient means for coeexpression of the two genes within individual cells. We then used the *lacZ*-IRES-*v-src* vector (LZIS) to test this approach *in vivo*. Embryos were injected with recombinant virions and subsequently analyzed as in the lineage experiments described above. Phenotypes of cells marked by LZIS were compared with those of cells marked by the lineage tracing virus, LZ10, which encodes only lacZ. In both skin and mesenchyme,

Table 2. Coexpression of two genes from single retroviral genomes. Data are from Ghattas et al. (1991)

	LacZ and CAT in individual infected cells (% of stained cells)		
	CAT only	LacZ only	LacZ plus CAT
Internal promoter – RSV	25	69	6
CMV	34	26	40
Regulated splicing	3	90	7
IRES – Types 1	3	0	97
Type 2	7	3	90
Type 3	2	2	96

[a] In each vector, the upstream gene was lacZ and the downstream gene was CAT (see Fig. 4). RSV, Rous sarcoma virus U3 segment; CMV, cytomegalovirus promoter.

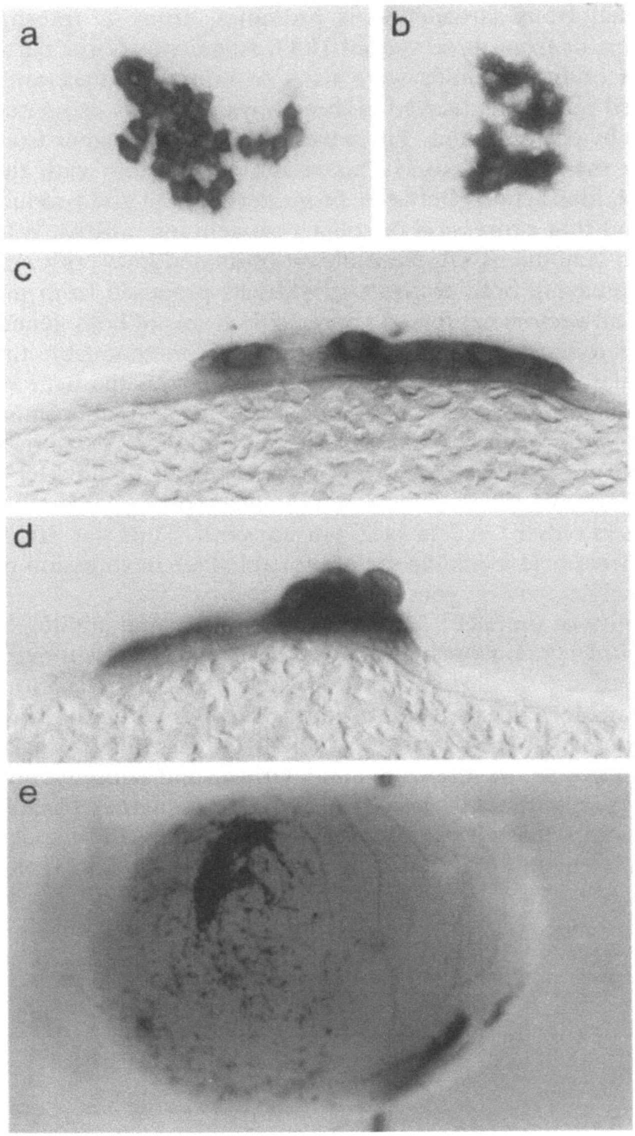

Fig. 5a–e. Cells in skin (a–d) and meninges (e) infected with lacZ (a, c) or lacZ-plus-v-src bearing vectors. Expression of v-src leads to tissue-specific morphological abnormalities that can be detected by means of the coexpressed marker, lacZ, a, b and e are whole mounts of individual clones; c and d are sections of plastic-embedded material. Reprinted from Ghattas et al. 1991.

lacZ-positive cells marked by LZIS displayed morphological abnormalities that were attributable to expression of v-src. Examples are shown in Figure 5. In skin, LZ10-marked clones were composed of flat polygonal cells with large central nuclei and scant cytoplasm (Fig. 5a; see also Galileo et al. 1990). In contrast, cells in LZIS-marked clones were large, irregularly shaped, and contained multiple inclusions (Fig. 5b). These cells were poorly integrated into the epidermis and frequently seemed to be delaminating from its apical surface (Figs. 5c and d). In addition, the average number of cells per clone was greatly reduced when v-src was present. Such abnormalities were seen in more than 90% of the clones analyzed, indicating that v-src was expressed in many (and perhaps all(lacZ-positive cells. Interestingly, the effects of v-src appeared to be tissue-specific; whereas expression of the oncogene appeared to block the normal proliferation of epidermal cells, it led to loss of proliferative control in the mesoderm, resulting in the production of large clones in meninges that ensheathe the brain and spinal cord, and in the loose mesenchyme that overlies the meninges (Fig. 5e). No comparably large clones of meningeal or mesenchymal cells were seen in LZ10-injected embryos. Although our aim is not to study the biology of v-src but rather to document the incidence of detectable abnormalities, it is worth noting that the morphological abnormalities we saw may be useful in studying cell-autonomous and tissue-specific effects of oncogene overexpression on individual cells in complex tissues.

Retrovirus-mediated transfer of antisense sequences

In many cases it would be more useful to remove a single gene from a cell than to overexpress that gene. A promising means to this end is to use vectors that produce antisense copies of mRNAs. Although the mechanisms of antisense inhibition of gene expression remain poorly understood and are clearly complex, there are now an increasing number of cases in which this approach has been used to delete a single gene product from the cell's repertoire with consequent effects on that cell's behavior (Colman 1990). To apply this method to the tectum, we wanted to construct retroviral vectors that encode lacZ plus an antisense copy of a gene whose product is implicated in cell migration. As an initial test of this strategy we used the integrin β1 gene. The integrins comprise a large family of genes that serve as cellular receptors for a variety of adhesive ligands in the extracellular matrix and on other cellular surfaces. Integrins are heterodimers of α- and β-subunits; the β-subunit is the major β-subunit expressed in the nervous system and is the β-1 subunit shared by a series of at least eight α-subunits, many of which are also expressed in the nervous system. Integrins have been implicated in the migration of neural crest cells and in the promotion of neurite outgrowth by neurons (Sanes 1989b). Furthermore, the β1-subunit is present on many cell types in the developing optic tectum during the period when cell migrations are occurring.

For all of these reasons, the integrin β1-subunit seemed to be a suitable candidate gene for initial tests of the antisense strategy. We therefore constructed vectors in which the lineage-tracing retrovirus was modified to contain antisense sequences for the chicken β1 integrin subunit. This virus (LZ15) can significantly inhibit (~ 60%) expression of integrin on cell surfaces, as judged by measurements on cultured cells. LZ15 or its parent (LZ10) was injected into E3 tecta and the resulting clones were compared at later stages. Preliminary results are summarized in Figure 6. At E6, when the tectum was still almost entirely a ventricular zone, both LZ10 and LZ15-marked clones consisted of radially arrayed cells extending across its width. Neither the size nor the shape of LZ15-marked clones was significantly affected by the presence of antisense sequences. By E7.5, however, the antisense sequences had begun to exert an effect. At this stage, a tectal plate surmounts the ventricular zone and the radial migration described above is underway. In LZ15-marked clones, cells remained radially arrayed, as in controls, but many accumulated in the ventricular zone and failed to enter the tectal plate. This perturbation remained evident at E9, the latest stage that we have examined so far. By E9, the tectal plate had thickened considerably as a result of cellular migration, and most cells in LZ10-marked clones (68% on average) lay within it. In contrast, only 22% of cells in the LZ15 marked clones were found in the tectal plate; most

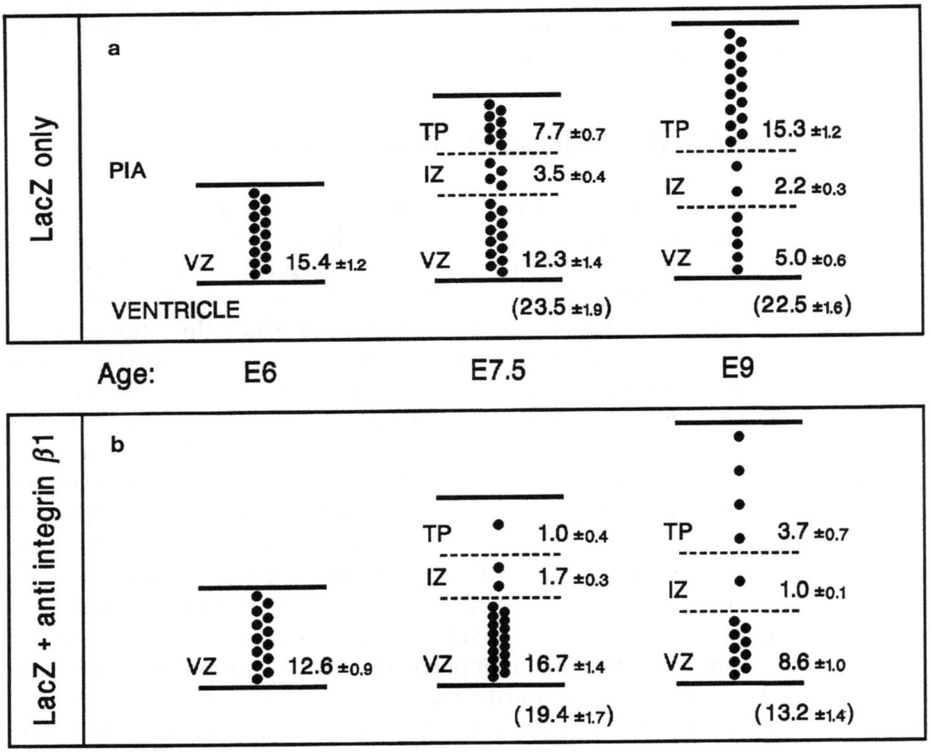

Fig. 6a, b. Perturbation of radial migration in optic tectum by introduction of sequences complementary to the intregrin β1 mRNA. Numbers are mean ± S.E. cells per lamina or per clone for each of the three ages analyzed. Either lacZ only (LZ10, A) or lacZ anti-integrin LZ15, B) virus was injected at E3. 30–60 clones were analyzed at each age for each vector.

remained in the ventricular zone. Furthermore, whereas LZ15-marked clones contained the same mean number of cells as LZ10-marked clones on E6 and E7.5, the LZ15-marked clones contained around 50% fewer cells than LZ10-marked clones by E9.

From these preliminary results (Galileo et al. 1991) we draw the following tentative conclusions. First, a reduction in levels of β1 integrin does not detectably perturb the radial stacking of clonally related cells in the ventricular zone. It does, however, lead to a blockade of radial migration along radial glia. Whether the crucial integrin molecules are those on the migrating cells, on the radial glial cells, or on both remains to be determined. In any event, it appears that cells deprived of sufficient quantities of integrins fail to migrate radially and instead accumulate in the ventricular zone. The subsequent loss of marked cells from LZ15-marked clones raises the possibility that those cells unable to migrate are unable to survive. Further studies will be needed to determine the fate of the cells whose migration has been blocked to define more precisely the stage at which migration is perturbed, and to learn whether expression of the antisense constructs affects the tangentially migrating cells that normally migrate along axons and differentiate into multipolar neurons.

In summary, our results raise the possibility that integrins, which had not previously been implicated in radial glial functin (Hatten 1990), may be involved in the migration of neuroblasts along radial glial cells in optic tectum. Moreover, the very fact that we can detect abnormal behavior in antisense-expressing cells encourages us to construct and test other vectors that target other adhesion molecules implicated in neuron-neuron and neuron-glia interactions.

Conclusions

Our development studies of optic tectum began with a description of genealogical relationship among cells. These studies demonstrated the existence of a multipotential tectal progenitor that gives rise to several neuronal and glial subtypes. We then extended these studies to map the migratory paths that clonally related cells take as they move from their birthplaces to their destinations. In so doing, we documented an intriguing relationship between migratory path and phenotypic choice. We are now trying to test whether this relationship is a causal one by producing small clones of transgenic cells whose migration and differentiation can be assessed in an otherwise unperturbed environment. Although these relationships are likely to become complex, it is our hope that two-gene and antisense vectors of the sorts we have constructed will be useful in unraveling them.

Acknowledgments. This work was supported by grants from NIH and the McKnight Foundation.

References

Alvarez-Buylla A, Theelen M, Nottebohm F (1990) Proliferation "hot-spots" in adult avian ventricular zone reveal radial cell division. Neuron 5:101–109

Arrigo S, Beemon K (1988) Regulation of Rous sarcome virus RNA splicing and stability. Mol Cell Biol 8:4858–4867

Bronner-Fraser M, Fraser S (1989) Developmental potential of avian trunk neural crest cells in situ. Neuron 3:755–766

Bunge MB, Wood PM, Tynan LB, Bates ML, Sanes JR (1989) Perineurium originates from fibroblasts: demonstration with a retroviral marker. Science 243:229–231

Colman A (1990) Antisense strategies in cell and developmental biology. J Cell Sci 97:399–409

Emmerman M, Temin HM (1986) Genes with promoters in retrovirus vectors can be independently suppressed by an epigenetic mechanism. Cell 95:459–467

Frank E, Sanes JR (1991) Lineage of neurons and glia in chick dorsal root ganglia: analysis *in vivo* with a recombinant retrovirus. Development 111:895–908

Frederiksen K, McKay RDG (1988) Profileration and differentiation of rat neuroepithelial precursor cells in vivo. J Neurosci 8:1144–1151

Friedman T (1989) Progress toward human gene therapy. Science 244:–1275–1281

Galileo DS, Gray GE, Owens GC, Majors J, Sanes JR (1990) Neurons and glia arise from a common progenitor in chicken optic tectum: demonstration with two retroviruses and cell type-specific antibodies. Proc Natl Acad Sci USA 87:458–462

Galileo DS, Majors J, Sanes JR (1991) Retrovirus-mediated introduction of antisense integrin β1 halts radial migration in chick optic tectum in vivo. Neurosci Abstr 17:923

Ghattas IR, Sanes JR, Majors JE (1991) The encephalomyocarditis virus internal ribosome entry site allows efficient coexpression of two genes from a recombinant provirus in cultured cells and in embryos. Mol Cell Biol, 11:5848–5859

Gray GE, Sanes JR (1991) Migratory paths and phenotypic choices of clonally related cells in the avian optic tectum. Neuron 6:211–225

Gray GE, Sanes JR (1992) Lineage of radial glia in the chicken optic tectum. Development, 114:271–283

Gray GE, Glover JC, Majors J, Sanes JR (1988) Radial arrangement of clonally related cells in the chicken optic tectum: lineage analysis with a recombinant retrovirus. Proc Natl Acad Sci USA: 85:7356–7360

Gray GE, Leber SM, Sanes JR (1990) Migratory patterns of clonally related cells in the developing central nervous system. Experientia 46:929–940

Hatten ME (1990) Riding the glial monorail: a common mechanism for glialguided neuronal migration in different regions of the developing brain. Trends Neurosci 13:179–184

Holt CE, Bertsch TW, Ellis HM, Harris WA (1988) Cellular determination in the Xenopus retina is independent of lineage and birth date. Neuron 1:15–26

Jackson RJ, Howell MT, Kaminski A (1990) The novel mechanism of initiation of picornavirus RNA translation. Trends Biochem sci 15:477–483

Jassik-Gerschenfeld D, Hardy O (1984) The avian optic tectum: neurophysiology and behavioral correlations. In: Vanegas H (ed) Comparative anatomy of the optic tectum Plenum Press; New York, pp 649–686

LaVail JH, Cowan WM (1971) The development of the chick optic tectum. II. Autoradiographic studies. Brain Res 28:421–441

Leber SM, Breedlove SM, Sanes JR (1990) Lineage, arrangement and death of clonally related neurons in chick spinal cord. J Neurosci 10:2451–2462

Luskin MB, Pearlman AL, Sanes JR 1988) Cell lineage in the cerebral cortex of the mouse studied in vivo and in vitro with a recombinant retrovirus. Neuron 1:635–647

Miller AD, Rosman GJ (1989) Improved retroviral vectors for gene transfer and expression. Bio Techniques 9:980–990

Price J, Thurlow L (1988) Cell lineage in the rat cerebral cortex: a study using retroviral-mediated gene transfer. Development 104:473–482

Rakic P (1988) Specification of cerebral cortical areas. Science 241:170–176

Sanes JR (1989a) Extracellular matrix molecules that influence neural development. Ann Rev Neurosci 12:521–546

Sanes JR (1989b) Analyzing cell lineage with a recombinant retrovirus. Trends Neurosci 12:21–28
Sanes JR, Rubinstein JCR, Nicolas JF (1986) Use of a recombinant retrovirus to study post-implantation cell lineage in mouse embryos. EMBO J 5:3133–3142
Stoltzfus CM, Fogarty SJ (1989) Multiple regions in the Rous sarcoma virus src gene intron act in cis to affect the accumulation of unspliced RNA. J Virol 63:1669–1676
Turner DL, Cepko CL (1987) A common progenitor for neurons and glia persists in rat retina late in development. Nature 328:131–136
Turner DL, Snyder EY, Cepko L (1990) Lineage-independent determination of cell type in the embryonic mouse retina. Neuron 4:833–845
Vanselow J, Thanos S, Godement P, Henke-Fahle S, Bonhoeffer F (1989) Spatial arrangement of radial glia and ingrowing retinal axons in the chick optic tectum during development. Dev Brain Res 45:15–27
Voigt T (1989) Development of glial cells in the cerebral wall of ferrets: direct tracing of their transformation from radial glia into astrocytes. J Comp Neurol 289:74–88
Walsh C, Cepko CL (1988) Clonally related cortical cells show several migration patterns. Science 241:1342–1345
Weiss R, Teich N, Varms H, Coffin J (1985) RNA tumor viruses. Gold Spring Harbor, Cold Spring Harbor Lab
Wetts R, Fraser SE (1988) Multipotent precursors can give rise to all major cell types of the frog retina. Science 239:1142–1145

Immortal Mammalian Neuronal Stem Cells Differentiate after Implantation into the Developing Brain

R. D. G. McKay

There is overwhelming evidence that neurons are derived from a multipotential precursor (McKay 1989). The first data on this subject were obtained as part of a series of pioneering experiments in the laboratory of Seymour Benzer at Caltech (Ready et al. 1976). In these studies the progeny of individual neuronal precursor cells were identified by an induced genetic change which allowed microscopic identification of cell clones in the developing retina. The cell types present in the marked clones were then identified. All neuronal and non-neuronal cell types in the Drosophila retina were found in marked clones without any bias. These results showed that in this tissue there is a multipotential precursor for neurons and non-neuronal cells. This conclusion has been substantiated and extended by a large number of studies that have identified genes that function in the pathways that define cell types in the Drosophila retina (Van Vactor et al. 1991; Simon et al. 1991). Transplants of neuroepithelial cells also show that local extracellular signals also regulate the fate of the multipotential neuronal precursors in mammals. A crucial signalling step that commits the multipotential precursor cell to a specific neuronal fate occurs in the few hours following the final round of DNA synthesis (McConnell 1988a, 1989, 1991). These experiments show that in vertebrates and invertebrates the switch from precursor to differentiated neuron is important in regulating the number and type of neuron found in the adult brain. This paper reviews work from our group on the identification and immortalization of multipotential mammalian neuronal stem cells. The relevance of these studies to cell therapy is discussed.

In vivo studies show that mammalian neuromal precursor cells can be identified by the specific expression of the intermediate filament protein, nestin (Frederiksen and McRay 1988; Lendahl et al. 1990). Nestin-positive neuroepithelial stem cells proliferate in response to defined growth factors *in vitro*. When the mitogenic growth factors are withdrawn, nestin is repressed, neurofilament is induced and the cells acquire neuronal morphology (Cattaneo and McKay 1990). These experiments show that the transition from precursor to neuron, measured by the expression pattern of specific intermediate filament proteins, occurs in tissue culture. The proliferation of nestin-positive cells in culture permits their infection by retrovirus. Nestin-positive cell lines can be readily generated by retrovirus transduction of immortalizing oncogenes (Frederiksen et al. 1988). These cell lines are stable for long periods in culture.

After transplantation, immortal precursor cells can differentiate into neurons and glia. The type of neuron generated depends on the site and developmental stage of the host tissue, suggesting that extracellular signals, derived from the host, regulate fate choice in the immoral cells (Renfranz et al. 1991).

The experiments reviewed here focus on two areas of stem cell biology:
1. developmental regulation of nestin;
2. factors which determine fate choice of immortal stem cells.

Nestin-positive cells differentiate into neurons

To identify neuronal stem cells monoclonal antibodies against the embryonic brain were screened by immunohistochemistry allowing identification of antigens expressed in the major morphologically distinct cell types in the developing brain (Hackfield and McKay 1985). Three major classes of cell were found: neurons, endothelial cells and radial glial cells. Antigens expressed in radial glial cells were the only candidates for antigens expressed in neuronal precursors, as both the neuronal and endothelial antigens were expressed too late to be present in neuronal precursors. In the central nervous system (CNS) neurons differentiate on a very precise schedule. The first post-mitotic neurons are the motor neurons of the spinal cord. In the rat, motor neurons differentiate from precursors on embryonic day 12 (E12). Neuron formation continues in the spinal cord for the next three days of development and the ceases. Measuring development permitted formal proof that the antigen recognized by the monoclonal antibody Rat 401 was expressed in neuronal precursor cells (Frederiksen and McKay 1988). The antigen carrying the epitope recognized by the Rat 401 antibody was cloned by screening an expression library of cDNAs from embryonic rat CNS and shown to be a new class of intermediate filament protein (Lendahl et al. 1990). The expression pattern of the complementary mRNA was identical to the pattern of expression of the original epitope mapped by the Rat 401 antibody. Antibodies raised against the bacterially expressed protein encoded by the cDNA recognized the same protein and cells as the original monoclonal antibody. These results prove that a large class of neuronal precursor cells transcribes a gene which is down-regulated as the neurons differentiate. This new intermediate filament protein was named nestin because it is specifically expressed in *ne*uroepithelial *st*em cells.

Evidence from primary cell culture experiments confirms that nestin is expressed in neuronal stem cells (Cattaneo and McKay 1990). More than 95% of the cells dissociated from the striatum of the rat on ernbryonic day 14 (E14) are nestin-positive. When these cells are placed in serum-free cultures containing mitogens, the nestin-positive cells profilerate, giving clones of nestin-positive cells. These cells proliferate for at least nine days in culture. If the mitogenic growth factors are then withdrawn, the cells change their morphology, extending long processes nestin expression is repressed and neurofilaments, the neuron-specific intermediate filament genes, are induced. These data provide direct proof for the in vitro transition from a nestin-positive precursor to a neurofilament positive product.

The third line of evidence that neuronal precursor cells are nestin-positive is provided by the properties of immortal cell lines derived from the brain. Nestin-positive cell lines can be derived from the embryonic nervous system by infection with retroviruses transducing oncogenes. Even though nestin expression is transient in vivo, these cell lines express nestin for long periods in culture. This result, like the data from primary cell culture, suggests that the precursor state is not inherently transient. Immortal nestin-positive cells can be induced to differentiate in culture (Frederiksen et al. 1988), but this process is very inefficient at present. In contrast, when immortal precursor cells are transplanted into the developing brain, they differentiate efficiently into neurons and glial cells (Renfranz et al. 1991). When 70000 cells were transplanted into either the hippocampus or cerebellum, many tens of thousands of transplanted cells acquire specific neuronal and glial morphologies.

The evidence derived from these three approaches – antigen expression in vivo, differentiation of primary cells in vitro and transplants of immortal cells – all points to the same conclusion: neurons are derived from nestin-positive precursors.

Local signals commit immortal nestin-positive cells to specific fates

Recombinant retroviruses have been used to immortalize cell lines expressing markers found in NEP cells

Nest+ cell lines have been made by infection of primary, embroynic brain cells with retroviruses carrying oncogenes. Cell lines were first generated by infecting cerebellar cultures with three retroviruses carrying three different oncogenes (Frederiksen et al. 1988). The *neu* oncogene gave very slowly growing lines. The *v-myc* and SV40 T-antigen oncogenes were much more efficient in establishing cell lines. Some of these cell lines were made with the temperature-sensitive T-antigen, tsA58, from SV40 virus. Growing the cells at the permissive or non-permissive temperature for T-antigen influenced their differentiated state (Hayes et al. 1991; Almazan and McKay 1992). This success encouraged us to make additional cells lines from hippocampus, striatum, retina, optic nerve, cerebral cortex and skin. The most successful differentiation has been seen after transplantation of hippocampal cells immortalized with T-antigen carrying both the tsA58 mutation and the U19 mutations (Renfranz et al. 1991). The ability to make cell lines from the embryonic brain is now well established and our major effort has switched to the functional analysis of these cell lines.

Immortal cell lines can function in the major developmental processes of neurogenesis and differentiate into functionally specialized neurons

In tissue culture the differentiation of immortal cells into neurons does occur (Frederiksen et al. 1988) but is inefficient. In contrast to neurons, immortal oligodendrocyte precursors undergo substantial differentiation in culture (Almazan and McKay 1992). Because the signals driving neuronal precursors to neuronal fates are only partly known, transplantation into the developing brains is the most rigorous test for the differentiation properties of conditionally immortal cells. To transplant immortal cells into regions of the newborn rath where neurons are still being generated, a stereotaxic transplantation procedure was developed to precisely place immortal cells into the developing brain. We have shown that cerebellar and hippocampal cell lines become integrated into the host tissue after implantation (Renfranz et al. 1991; McKay et al. 1990). When a hippocampal cell line, HiB5, wis transplanted into new born hippocampus or cerebellum, the immortal cells stop dividing and differentiate into neurons and glia. In contrast, an established fibroblast line continues to divide and shows no sign of neuronal differentiation (Renfranz et al. 1991). HiB5 cells were derived from embryonic (E16) hippocampus cells and, when transplanted into newborn hippocampus, the cells acquire the morphology of dentate gyrus granule neurons (DGNs). Many interesting questions can be addressed by transplanting immortal NEP cells which are still competent to differentiate into adult fates:

Can a clonal cell give rise to different kinds of neuron?

In contrast to their behavior in developing hippocampus, HiB5 cells acquire morphological features of cerebellar granule neurons (CGN) when they are placed in the developing cerebellum (Renfranz et al. 1991). The cell bodies of the transplanted cells are co-distributed with the cell bodies of the endogenous granule neurons in the granular layer. Retrograde axonal labelling and florescent tagging procedures show that transplanted cells can extend axons and dendrites with morphologies identical to normal granule neurons.

Antigen expression has been used to further define the differentiation of immortal, transplanted cells. In these experiments the transplanted cells are marked with fluorescent latex beads and cell type-specific antigen expression determined by immunohistochemistry. Neuron-specific enolase (NSE) is expressed in HiB5 cells transplanted to both hippocampus and cerebellum. Calbindin is a calcium binding protein expressed in many DGNs but not CGNs (Feldmann and Christakos 1983). Confocal microscopy shows that calbindin is expressed in transplanted fluorescein-bead labelled HiB5 cells in the granule cell layer of the hippocampus, while HiB5 cells in the cerebellar granule cell layer do not express calbindin. This result suggests that transplanted cells acquire specific biochemical characteristics of DGNs in response to local environmental cues.

Can the same cell give rise to neurons and glia?

The intermediate filament GFAP is specifically induced in astrocytes, and HiB5 cells expressing GFAP are readily found in transplanted hosts. To be certain that the original cell line was clonal, independent subclones of HiB5 were used in these experiments. Subclones were picked after a second round of retrovirus infection. The clonal nature of the subclones was established by southern blots, which showed independent integration sites following the second round of retro-infection. These results showed that a clonal cell can efficiently differentiate into both neurons and glia.

Are the differentiated neurons functional? The initial approach we are taking to this question is to use induction of the immediate early response gene *c-fos* to identify DGNs derived from HiB5 which are activated in pharmacologically induced seizures. The conditions for this experiment have been defined by others in adult animals (Sheng and Greenberg 1990; Dragounow and Robertson 1987; Dragounow et al. 1989). The technical limit on this experiment is the ease of identification of a transplanted neuron in a slice preparation of hippocampus. Fluorescent-bead and 3H-thymidine labelling mark transplanted cells in 10-week-old rats transplanted at post-natal day 2 (P2). *c-fos* is strongly induced in PTZ or kainic acid-treated adult rat DGNs (Smeyne et al. 1992) and clear induction of *c-fos* can be seen in 3H-thymidine labelled cells integrated into the granule cell layer of the hippocampal dentate gyrus. These findings suggest that many of the transplanted cells are functionally integrated into the brain.

Although *c-fos* induction is a useful assay for the physiological response of transplanted cells, a more detailed electrophysiological analysis would be very interesting. Electrophysiological analysis of transplanted cells would be made simpler by a strong quantitative marker allowing the identification of living transplanted cells in an acute slice from an adult rat brain. To quantitatively identify transplanted HiB5 cells we are transfecting reporter genes (lacZ, alkaline phosphatase and luciferase) into HiB5 cells under the control of constitutive and cell type specific promoters. lacZ and luciferase expression would allow living transplanted cells to be identified in electrophysiological experiments. The expression of alkaline phosphatase may not be compatible with the detection of living transplanted cells, as the background alkaline phosphatase activity has not been measured. However, the expression of surface bound alkaline phosphatase may be a third useful marker to follow transplanted cells.

Can genetic methods be used to dissect the fate and function of these cells?

As immortal cells can be genetically manipulated in tissue culture, a key advantage of transplanting an immortal cell is that a restricted number of genetically modified cells can be integrated into a normal (wild type) background. The transplanted cells are derived from a clonal precursor, considerably simplifying the interpretation of mechanisms, either developmental or functional, which depend on cell-cell interaction. To make full use of immortal NEP cells we must develop simple quantitative assays for cell differentiation. Quantitative assays will permit a systematic approach to the influence of transgenes on the differentiation and function of transplanted NEP cells. As a first step in this approach we have expressed the E. coli *lacZ* gene in HiB5 cells. Cells expressing *lacZ* were isolated by FACS analysis and transplanted. The X-gal histochemical reaction shows that cells transplanted into the cerebellum integrate into the host tissue (Renfranz et al. 1991).

Discussion

Our data show that precursor cell lines derived from embryonic hippocampus differentiate into neurons after transplantation into the newborn hippocampus. The same cell give rise to neurons appropriate to the site of transplantation. The type of neuron generated at each site is characteristic of the neurons that are being generated by the host tissue, by both morphological and specific gene expression assays. As the transplanted cells are truly clonal, these results show that immortal cells are multipotential and become recruited to specific fates by local signals. The ability of immortal stem cells to differentiate allows the mechanisms of multipotentialy to be explored with new tools.

Transplantation of appropriate cell types is a potential therapeutic strategy for neurodegenerative disease (for extensive reviews see TINS, August 1991). Much of the attention focussed on transplantation is a consequence of the extraordinary potential of transplanted embryonic tissue to differentiate into neurons, extend axons and form synapses in the adult host brain (Björklund et al. 1976). In animal models, transplants of embryonic tissue reproducibly give behavioral recovery from experimentally induced lesions. Perhaps Parkinson's disease is the most likely disease to respond to this approach, because the neuronal deficit is so specific and preliminary results in Parkinsonian patients are now encouraging (Lindvall et al. 1989, 1990); Widner et al. 1991). However, transplantation may also be a useful approach to aspects of other CNS diseases, including Alzheimer's and Huntington's diseases.

This paper has focussed on two results from our lab. First, that neuronal precursor cells specifically express the intermediate filament protein nestin. Second, that nestin-positive cell lines can be generated from the developing brain and retain the ability to differentiate into neurons and glia. These results may have important implications for cell therapy in mammals.

Identification of neuronal precursors in man

The identification of a gene which is expressed in multipotential neuronal precursors may allow this cell type to be characterized in the developing human nervous systems. We have expressed the nestin coding sequences in bacteria and raised polyclonal antisera against the fusion protein. These antisera recognize nestin from many species not recognized by the original monoclonal antibody. In particular, we have shown that the nestin gene is expressed in CNS stem cells of mouse and man (Valtz et al. 1991; Tohyama et al. 1992). The expression pattern of nestin in developing human CNS is very similar to the expression pattern in rat and mouse, suggesting that anti-nestin antibodies may define a human multipotential precursor cell type.

When do precursor cells become committed to specific neuronal fates?

If a precursor cell is multipotential, when does a neuron become committed to a specific fate? This question has been addressed in a series of transplant experiments on developing cerebral cortex of the ferret (McConnell 1988, 1991). In the cortex of all mammals, neurons with distinct functions are found in different layers. It has been known for some time that neurons in different layers become post-mitotic in a very precise sequence, suggesting that the time when a cell becomes post-mitotic determines neuronal type. The results of the ferret cortical transplants define cells type commitment as occurring in a very short period of a few hours following the last S phase of the precursor. If precursor cells are transplanted immediately following labelling with 3H-thymidine, they are converted to the fate appropriate to the host tissue. But if they are allowed to remain at the donor site for as short a period as four hours after the S phase defined by the thymidine injection, then cells are committed to the fate appropriate to the donor site. The cells which are committed are those that become post-mitotic; cells that go through another round of DNA synthesis in the host convert to the fate expected of host cells. These results suggest that precursor cells differentiate in response to local signals which are present in the ventricular zone of the developing germinal epithelium.

Primary and immortal cells have different uses in transplantation

There are many different potential approaches to cell therapy in the central nervous system. There are several reports of the delivery of growth factors via transplanted established fibroblasts and, at least in short-term experiments, these transplanted cells significantly alter the properties of the host animal at the cellular level. In most cases established cell lines continue to grow after transplantation. Primary cells overcome the problem of continued growth. Primary fibroblasts and primary astrocytes are potentially useful donors, as they can be obtained in large numbers and, in principle, they can be genetically manipulated to express therapeutically useful gene products. From the perspective of clinical application there is a second major advantage: it may be possible to obtain these cell from the intended host, thus reducing problems of immune rejection. An important technical issue in these experiments is the long-term expression of transgenes in transplanted cells. It may be difficult to obtain high levels of gene expression from introduced genes following transplantation.

Immortal precursor cell lines which differentiate into neurons and astrocytes might be another useful strategy for cell therapy. The advantages of cell lines are discussed in detail Lendahl and McKay (1990). Here we discuss a major potential disadvantage. If cell lines are to be useful in this way, then some fundamental questions in the biology of neural development must be answered.

Immortalizing steps must be comptabile with long-term controlled cell survival in the host and perform a useful function in the host brain.

Immortalization and Tumor Formation

During normal development nestin is expressed in neuronal precursor cells and very few cells are nestin positive in the differentiated brain. We have shown that experimentally immortalized cells can express nestin for long periods in culture. These results raise the question wihether cells that are transformed by the processes that generate CNS tumors also express nestin?

Antisera against nestin have been used to characterize primary human tumors and cell lines derived from human tumors (Valtz et al. 1991; Tohyama et al. 1992). Most information on nestin expression is currently available on a tumor type which occur predominantly in young children. These tumors are called medulloblastomas because they contain several differentiated cell types, most commonly neurons and glia and less frequently muscle cells. Because of this the tumors were thought to be derived from a multipotential stem cell, which must be self renewing and still capable of differentiation. The hypothetical stem cell was called a medulloblast, and the tumor, medulloblastoma, because these tumors are often found in the midline of the hindbrain close to the medulla and cerebellum.

Very large numbers of cerebellar granule neurons are generated late in brain development. The large numbers and late development of these cerebellar neurons make it possible that the medulloblast is derived from their precursors. Nestin-positive cells are present in developing cerebellum and nestin-positive cell lines can give rise to cerebellar neurons. The nestin-positive cell line ST15A was derived from the developing cerebellum and can differentiate in culture to neuronal, glial or muscle fates (Valtz et al. 1991). This result shows that a clonal cell can give rise to three differentiated cell types found in medulloblastoma and is consistent with the medulloblast hypothesis.

Primary medulloblastoma tumors and cell lines derived from medulloblastomas contain large numbers of nestin-positive cells (Valtz et al. 1991; Tohyama et al. 1992). These results are consistent with a nestin-positive, multipotential renewing stem cell giving rise to the medulloblastoma. However, nestin-positive cells are also associated with astrocytomas and glioblastomas, raising the possibility that tumor cells derived from the brain express nestin and that nestin expression is not a specific feature of medulloblastomas.

Although our results show that human CNS tumors contain nestin-positive cells, it should be stressed that nestin-positive cell lines do not always form tumors following transplantation. We have a great deal of experience showing that cells conditionally immortalized with a temperature-sensitive allele of SV40T-antigen do not show abnormal growth in the newborn rat hippocampus or cerebellum. In contrast, established fibroblast line show unregulated growth which is detectable within a few days following transplantation. These results suggest that cells immortalized in vitro may be useful in cell therapy, but of course a more detailed analysis of growth regulation would first be required.

For direct applications of immortal cells in cell therapy, it will be particularly important to understand the mechanisms which normally swith the rapidly proliferating neursonal stem cell to a terminally differentiated state.

Studies on cell growth show that the CDC2 and cyclin gene products play critical roles in transit through the cell cycle. During brain development these genes are transcriptionally represed, coincident with terminal differentiation (Hayes et al. 1991). These results suggest that the terminal differentiation (Hayes et al. 1991). These results suggest that the terminal differentiation of neurons in the CNS may be regulated by mechanisms which are similar to those employed in the differentiation of yeast cells in response to extracellular, mating factor signals. Although little is currently known about the mechanisms which swich neuroepithelial stem cells from the proliferative to the differentiated state, the possibility that common motifs are employed in other, intensely studied systems suggests that a comprehensive picture of this process is possible. In the longer term, this knowledge may permit cell lines to be designed which are capable of differentiation into neurons and have well-understood mechanisms that limit growth.

References

Almazan G, McKay R (1992) An oligodendrocyte precursor cell line from rat optic nerve Brain Res 234–245

Bjorklund A, Stenevi U, Svengaard N (1976) Growth of transplanted monoaminergic neurones into the adult hippocampus along the perforant path. Nature 262:787–790

Cattaneo E, McKay RDG (1990) Proliferation and differentiation of neuronal stem cells regulated by nerve growth factor. Nature 347:762–765

Dragunow M, Robertson HA (1987) Kindling stimulation induces c-fos protein(s) in granule cells of the rat dentate gyrus. Nature 329:441–442

Dragunow M, Abraham WC, Goulding M, Mason SE, Robertson HA, Faull RL (1989) Long-term potentiation and the induction of c-fos mRNA and proteins in th dentate gyrus of unanesthetized rats. Neurosci Lett 101:274–280

Feldman SC, Christakos S (1983) Vitamin D-dependant calcium-binding protein in rat brain: biochemical and immunocytochemical characterization. Endocrinology 112:290–302

Frederiksen K, McKay RDG (1988) Proliferation and differentiation of rat neuroepithelial precursor cells in vivo. J Neurosci 8:1144–1151

Frederiksen K, Jat PS, Valtz N, Levy D, McKay RDG (1988) Immortalization of precursor cells from the mammalian CNS. Neuron 1:439–448

Hayes TE, Valtz NL, McKay RDG (1991) Downregulation of CDC2 upon terminal differentiation of neurons. New Biologist 3:259–269

Hockfield S, McKay RDG (1985) Identification of major cell classes in the developing mammalian nervous system. J Neurosci 5:3310–3328

Lendhal U, Zimmermann L, McKay RDG (1990) CNS stem cells express a new class of intermediate filament protein. Cell, 60:585–595

Lendhal U, McKay RDG (1990) The use of cell lines in neurobiology. Trends Neurosci 13:132–136

Lindvall O, Rehncrona S, Brundin P, Gustavii B, Astedt B, Widner H, Lindholm T, Bjorklund A, Leenders KL, Rothwell JC et al. (1989) Human fetal dopamine neurons grafted into the striatum in two patients with severe Parkinson's disease. A detailed account of methodology and a 6-month follow-up. Arch Neurol 46:615–631

Lindvall O, Brundin P, Widner H, Rehncrona S, Gustavii B, Frackowiak R, Leenders KL, Sawle G, Rothwell JC, Marsden CD, Bjorklund A (1990) Grafts of fetal dopamine neurons survive and improve motor function in Parkinson's disease. Science 247:574–577

McConnell SK (1988) Fates of visual cortical neurons in the ferret after isochronic and heterochronic transplantation. J Neurosci 8:945–974
McConnel SK (1989) The determination of neuronal fate in the cerebral cortex. Trends Neurosci :342–349
McConnel SK, Kaznowski CE (1991) Cell cycle dependence of laminar determination in developing neocortex. Science, 254:282–285
McKay RDG (1989) The origins of cellular diversity in the mammalian central nervous system. Cell 58:815–821
McKay RDG, Valtz V, Cunningham M, Hayes T (1990) Mechanisms regulating cell number and type in the mammalian central nervous system. Cold Spring Harb Symp Quant Biol 55:291–301
Ready DF, Hanson TE, Benzer S (1976) Development of the Drosophila retina, a neurocrystalline lattice. Dev Biol 53:217–240
Renfranz P, Cunningham M, McKay RDG (1991) Region-specific differentiation of the hippocampal stem cell line HiB5 upon implantation into the developing mammalian brain. Cell 66:713–729
Sheng M, Greenberg ME (1990) The regulation and function of c-fos and other immediate early genes in the nervous system. Neuron 4:477–485
Simon MA, Bowtell DD, Dodson GS, Laverty TR, Rubin GM (1991) Ras-1 and a putative guanine nucleotide exchange factor perform crucial steps in signaling by the sevenless protein tyrosine kinase. Cell 67:701–716
Smeyne RJ, Schilling K, Robertson L, Luk D, Oberdick J, Curran T, Morgan JI (1992) Fos-lacZ transgênic mice: mapping sites of gene induction in the central nervous system. Neuron 8:13–23
Tohyama T, Lee VM, Rorke LB, Marvin M, McKay RGD, Trojanowski JQ (1992) Nestin expression in embryonic human neuroepithelium and in human neuroepithelial tumor cells. Lab Invest 66:303–313
Valtz NL, Hayes TE, Norregaard T, Liu SM, McKay RDG (1991) An embryonic origin for medulloblastoma. New Biologist 3:364–371
Van Vactor DL Jr, Cagan RL, Kramer H, Zipursky SL (1991) Induction in the developing compound eye of Drosophila: multipe mechanisms restrict R7 induction to a single retinal precursor cell. Cell 67:1145–1155
Widner H, Brundin P, Rehncrona S, Gustavii B, Frackowiak R, Leenders KL, Sawle G, Rothwell JC, Marsden CD, Bjorklund A et al. (1991) Transplanted allogeneic fetal dopamine neurons survive and improve motor function idiopathic Parkinson's disease. Transplant Proc 23:793–795

Establishment of Clonal Cell Lines for the Study of Neural Function and Dysfunction

B. H. Wainer, D. N. Hammond, H. J. Lee, J. D. Roback, H. K. Choi, L. Won,* and *A. Heller*

Introduction

The use of gene transfer as a therapeutic approach to diseases of the nervous system depends on the assumption that neurological and psychiatric disorders are a result of defective gene expression and can be corrected by application of modern molecular approaches. The ultimate success of this strategy requires a detailed understanding of normal gene expression in the nervous system. In the case of Parkinson's disease (Fig. 1b), the primary therapeutic approach has been to either replace the dopamine pharmacologically (i.e., L-DOPA; Pletscher 1990), or to graft fetal dopaminergic neurons (Lindvall 1991). In the case of experimental Parkinsonism, genetically engineered cells have been grafted that synthesize and release dopamine (Gage et al. 1991). While this strategy may be effective for palliative purposes, it does not address the etiological and pathogenetic factors that actually initiate the degenerative process. In fact, it is quite possible that the grafted cells themselves could succumb to the basic disease process. In neurological disorders such as Alzheimer's disease (AD; Fig. 1a), a variety of neural pathways are selectively vulnerable and not all of the neurotransmitter deficits responsible for the devastating cognitive loss are understood (Saper et al. 1987; Katzman and Saitoh 1991; Decker and McGaugh 1991). For these reasons, "neurotransmitter" replacement strategies may not be as straightforward or efficacious as in the case of Parkinson's disease. While nerve growth factor (NGF) administration is being considered as a therapy to prevent cholinergic degeneration in AD (Phelps et al. 1989; Marx 1990), the fact that related substances, the neurotrophins (Maisonpierre et al. 1990), have recently been identified, and that other trophic factors may be involved in cholinergic neuronal viability, suggests that sucessful trophic therapy in AD may be complex (Barde 1989; Hefti et al. 1989). Since little information is available regarding putative trophic factors in the non-cholinergic pathways that are affected in AD, additional information regarding this disease process will be necessary before gene transfer or pharmacological approaches can be devised that are likely to be of much use. It is therefore evident that a major challenge to understanding and treating such diseases, either by pharmacologic or genetic approaches, is to determine the molecular

* The University of Chicago, Chicago IL 60637 USA

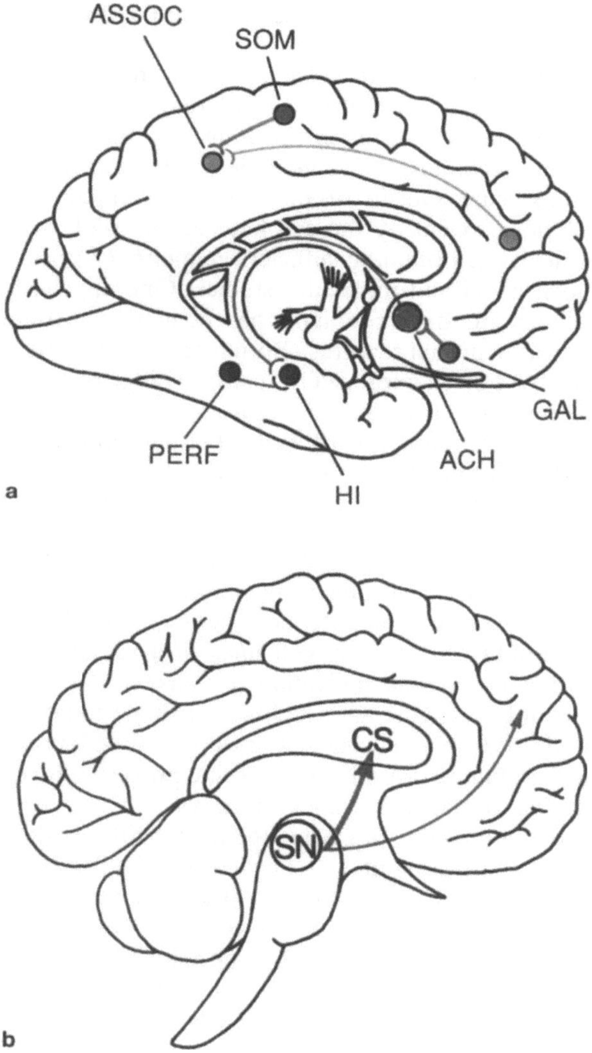

a

b

Fig. 1a, b. Schematic illustration of neural pathways that are selectively vulnerable in Alzheimer's disease. These pathways include: cortico-cortico association pathways (ASSOC), hippocampal circuitry (HI) including the perforant (PERF) pathway; diffuse ascending neuromodulatory systems, particularly the cholinergic (ACH), but may include dopaminergic and noradrenergic (not shown); and certain populations of interneurons containing somatostatin (SOM) and galanin (GAL). (b) Schematic illustration of central dopaminergic projections from the substantia nigra (SN) to the corpus striatum (CS) and cortical areas (thin red arrow). The degeneration of the nigrostrial projection (thick red arrow) plays a major role in the pathogenesis of Parkinson's disease.

mechanisms that mediate formation and maintenance of specific neural pathways (Purves 1988) and the processes that lead to dysfunction. With regard to pharmacologic approaches it is frequently the case that a particular drug may be efficacious because of its broad range of pharmacological effects rather than the exact mode of action which is presumed to be responsible for its therapeutic value. In the case of gene therapy, however, it is apparent that a successful result will depend heavily on a thorough understanding of the detailed pathophysiology of any particular disease process. This type of analysis in brain is made exceedingly difficult by the extensive heterogeneity of cell types and connections that may be affected even within a specific brain region or projection.

Given the difficulty of determining the etiology of complex diseases such as AD, it is unlikely that much progress can be made by using approaches which involve studies on the intact brain. For these reasons, a variety of primary cell culture techniques have been developed to provide simplified systems for study (Banker and Goslin 1991a). These techniques are, however, limited to a considerable extent by the cell heterogeneity of specific regions of brain and the difficulty of obtaining sufficient numbers of cells of a specific type for detailed biochemical and molecular studies. One approach that has been employed to circumvent these problems has been the utilization of clonal cell lines that exhibit neuronal features (Bottenstein 1981; Schubert 1984; Banker and Goslin 1991b). For example, PC12 cells, arising from a rat pheochromocytoma, have been used in a variety of studies to elucidate the response to and mechanisms of action of nerve growth factor (Tishler and Greene 1975; Greene et al. 1991). Although this approach has provided a considerable amount of information, it is limited for a number of reasons. Cell lines are derived from spontaneously arising tumors and therefore carry with them the inherently malignant nature of their cells of origin. While a recent report has described the isolation of a human neural cell line from non-neoplastic tissue (Ronnett et al. 1990), the frequency of such spontaneous "immortalization" events is extremely low. The possibility of obtaining cell lines of a specific neurochemical type from such spontaneous immortalization is unlikely. In addition, most of the cell lines currently available are subclones of single tumors of peripheral nervous system origin, i.e., pheochromocytomas or neuroblastomas, and are therefore of limited usefulness for studying specific processes involved in the development and survival of particular CNS pathways.

To date, there has been relatively little work carried out on the establishment of permanent cell lines from specific brain regions that elaborate or respond to trophic signals. The engineering of such cell lines can be approached from two points of view. One is the use of retroviral-mediated introduction (transduction) of oncogenes to "immortalize" primary brain cells (Lendahl and McKay 1990; Cepko 1989). This approach is useful for studying the properties of progenitor cells, but is probably of limited utility as an approach to the study of more mature differentiated cells. Retroviral transduction is possible only with mitotic cells, since the retroviral DNA can only be inserted into the host genome during replication. In addition, once a cell is "immortalized" it tends to remain locked within a particular developmental window. This latter phenomenon has been exploited by immunologists to study the stages of lymphocyte differentiation (Paige and Wu 1989; Alt et al. 1987). Given these limitations, viral gene transduction might yield cell lines for the study of early stages of neuronal development, but it is unlikely to provide cells that express the total phenotypic repertoire of mature neurons, which are invariably postmitotic.

A second approach has been to employ somatic cell fusion techniques in which primary brain cells are fused to a neuroblastoma cell line by exposure to polyethylene glycol (Hammond et al. 1986). This fusion technique permits one to "immortalize" postmitotic cell populations and is therefore more likely to yield cells which express highly differentiated neuronal phenotypes. In this chapter, the methodological considerations that are essential in applying the somatic cell fusion approach to the immortalization of CNS cells, and the properties of hybrid cell lines that have been generated from neurons of the septohippocampal and nigrostriatal pathways, will be discussed. The approach described here (i.e., the somatic cell fusion technique) makes available large numbers of brain region-specific clonal cells for cellular and molecular studies of specific neuronal circuits. The availability of such cells should serve as a useful adjunct to the detailed examination of the underlying molecular basis of central nervous system disease processes.

Methodology

Use of the somatic cell fusion technique to immortalize cells has been applied to a wide variety of cellular and genetic questions (Shay 1982). This is the method that has been used to generate lymphoid hybridoma cell lines which produce monoclonal antibodies (Kohler and Milstein 1975). In the case of the nervous system, the somatic cell fushion technique has lead to the development of several cell lines derived from sympathetic neurons (Greene et al. 1975) or dorsal root ganglion cells (Platika et al. 1985). We have exploited the cell fusion approach, illustrated schematically in Figure 2, because this technique enables one to immortalize immature central neurons as well as neurons that are postmitotic and therefore committed to a particular neurochemical phenotype and neuroanatomic pathway (Hammond et al. 1986, 1990; Lee et al. 1990a, b). This approach consists of fusing primary brain cells with N18TG2 neuroblastoma cells and growing the cell fusion products in a medium which selects for viable hybrid cells. The N18TG2 cells are deficient in the enzyme, hypoxanthine phosphoribosyltransferase (HPRT), and are therefore unable to utilize exogenous sources of hypoxanthine for purine synthesis. Following fusion, the cells are cultured in a medium containing hypoxanthine, aminopterin and thymidine (HAT). Since aminopterin blocks endogenous pathways required for purine and pyrimidine synthesis, the parent N18TG2 cells will not survive. Hybrid cells, however, should have the HPRT defect corrected by the inclusion of chromosomes from parental primary brain cells, thus enabling the hybrid cells to express functional enzyme and utilize exogenous sources of hypoxanthine.

The ability of a cell to grow in HAT medium strongly suggests it is a hybrid and has retained the replication characteristics of the parent neuroblastoma as well as the HPRT gene from the parent primary cell. However, since HPRT revertants can develop, despite a low frequency of occurrence (less than 10^{-7}), it is essential that the hybrid nature of a cell line also be confirmed. For this purpose, karyotype analysis and enzyme polymorphisms (Hammond et al. 1986) should be determined before evaluating the cell line for specific neurochemical properties.

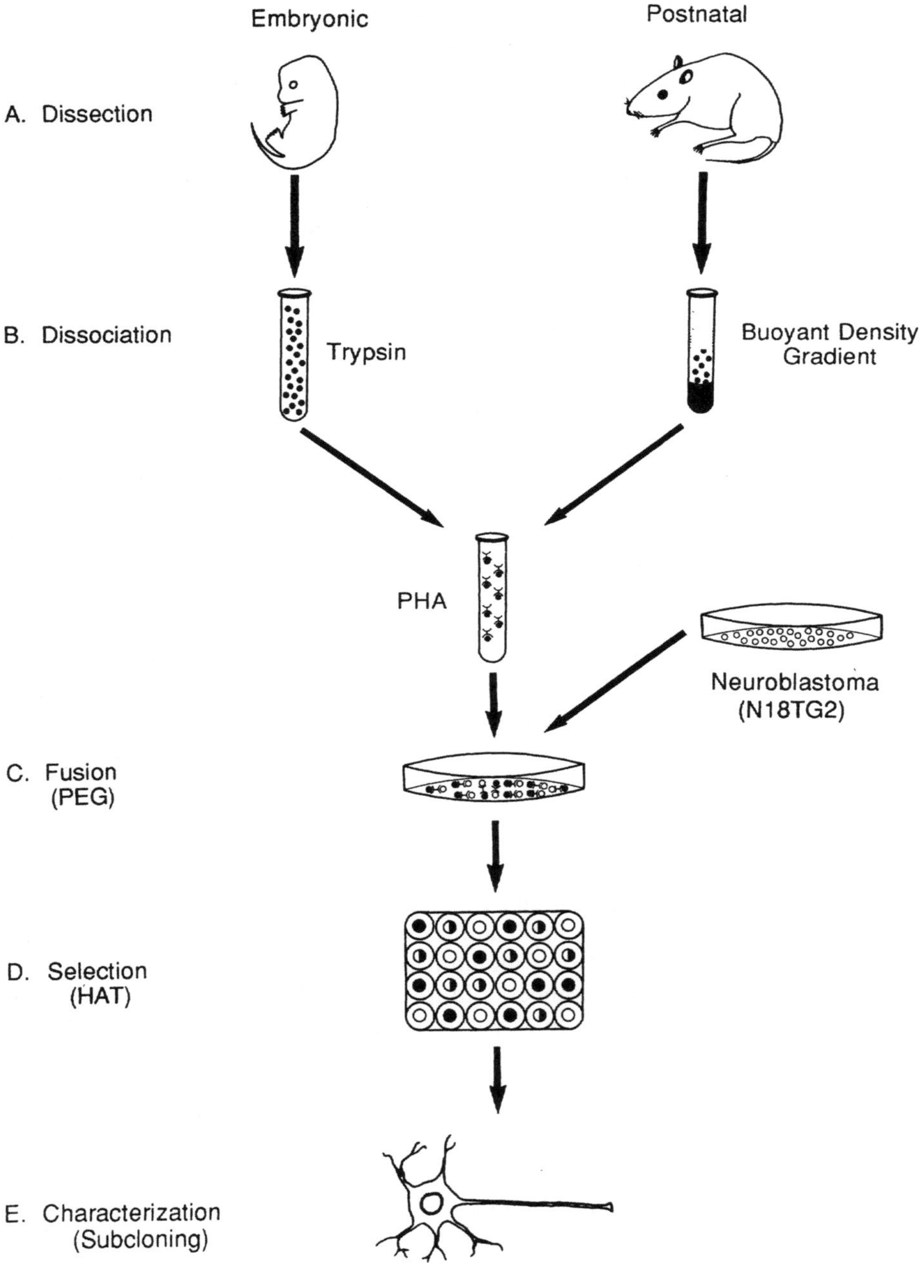

Fig. 2a–e. Schematic illustration of the generation of somatic hybrid cells. (a) Specific brain regions are dissected from embryonic or postnatal animals at an age when the neurons of interest have become postmitotic or have developed the appropriate neuronal phenotypes of interest. (b) Primary brain cell suspensions are obtained following trypsinization and mechanical dissociation. For harvesting of primary postnatal brain cells, a buoyant density gradient is employed. (c) The somatic cell fusion is conducted by placing the suspension of primary brain

Clonal Hybrid Cell Lines of the Septohippocampal Pathway

The septal region contains the rostral portion of the magnocellular basal forebrain system. This system, which projects diffusely to the cortex (Wainer and Mesulam 1990), plays an essential role in cognitive function (Nilsson et al. 1987; Delacour et al. 1990; Decker and McGaugh 1991), contains a substantial cholinergic component (Rye et al. 1984; Wainer et al. 1985) and is selectively vulnerable in AD (Fig. 1a; Katzman and Saitoh 1991; Decker and McGaugh 1991). The principal target of the septal cholinergic projection is the hippocampus (Wainer et al. 1985), which provides cholinergic trophic support through the elaboration of neurotrophins and other putative trophic factors (Kesslak et al. 1987; Gage et al. 1988; Barde 1989; Hefti et al. 1989). The septohippocampal pathway has been employed as a model system for the study of mechanisms of neuronal injury (Nilsson et al. 1987), plasticity (Cotman et al. 1973) and the effects of neuronal growth factors (Gage et al. 1988). The pathway is well defined anatomically but is composed of a complex series of cell types, and there is a growing literature with regard to trophic factors that influence this pathway, not only during development but in the aging process and following specific injury (Hefti et al. 1989; Fischer et al. 1987; Rosenberg et al. 1988; Knusel et al. 1991; Alderson et al. 1990). We have utilized the somatic cell fusion technique to immortalize neurons of the septal region and their target cells of the hippocampus as an approach to examining the trophic factors and responses involved in the development and survival of this essential limbic system pathway (Hammond et al. 1986, 1990; Lee et al. 1990).

To date, septal hybrid cell lines have been generated from both embryonic and young adult brains (Hammond et al. 1986, 1990; Lee et al. 1990b). We have studied in greatest detail those cells which were initially selected on the basis of their morphology. Colonies exhibiting reasonable generation times of 24–72 hours and showing "neurite-like processes" were isolated for expansion, subcloning and further analysis. More slowly growing colonies which demonstrated flat "glial-like" appearances were not examined in any detail.

The septal hybrid cell lines exhibit many properties of primary neurons (Table 1). The cells of interest have somata and processes which are highly refractile and, at the ultrastructural level, these processes display neuronal intermediate filaments and a number of other features of developing neurons, such as abundant growth cones (Hammond et al. 1990). The presence of

---

cells that have been treated with phytohemagglutinin (PHA) to enhance cell adhesion onto a monolayer culture of N18TG2 neuroblastoma cells and briefly exposing the two cell types to polyethylene glycol (PEG). (d) Microcultures are grown in HAT-containing media in which unfused neuroblastoma cells. Microcultures are grown in HAT-containing media in which unfused neuroblastoma cells (white) and primary brain cells (black) will exhibit no further growth, but fused hybrid cells (white/black) will replicate and colonies will be visible in 2–3 weeks. (e) Hybrid cell colonies are explanded, subcloned and screened for appropriate neurochemical markers.

Table 1. Properties of "cholinergic" hybrid septal cell lines[a]

Cytoskeleton:	NFP		+
	GFAP		−
Excitability:	Na^+-Currents and Action Potentials		+
Neurochemical profile:	Acetylcholine		+
	Choline acetyltransferase		+
	High affinity choline uptake		+
	K^+-stimulated ACh release		+
	Acetylcholinesterase		+
Growth factor response:	Neurotrophins:	NGF	−
		BDNF	(not tested)
	Cytokines:	IL3	+
		GM-CSF	+
Target vs. non-target cell recognition:			(not tested)
Synapse formation:			(+)[b]

[a] Abbreviations: NFP, neurofilament proteins; GFAP, glial fibrillary protein; ACH, acetlycholine; NGF, nerve growth factor; BDNF, brain derived neurotrophic factor; IL3, interleukin 3; GM-CSF, granulocyte macrophage colony stimulating factor.
[b] Preliminary observation.

neurofilaments has been demonstrated by immunocytochemical (Fig. 3A) and immunoblotting techniques. The cell lines demonstrate sufficiently differentiated cytoskeletal features such that they express both the low molecular weight forms of the neurofilaments as well as differentially phosphorylated isoforms of the middle and high molecular weight subunit neurofilaments. Such a cytoskeletal array is characteristic of a differentiated mature neuron (Lee et al. 1991; Carden et al. 1987). The cytoskeletal character of these lines is quite stable, and cells cultured continuously for over one year still express all of the neurofilament subunit isoforms and probably all of the various kinases and phosphatases required for the regulation of the appropriate neurofilament phosphorylation states. It should be noted that intermediate filaments characteristic of non-neuronal cells, such as cytokeratins or glial fibrillary acidic protein (GFAP), are not expressed in septal lines which contain neurofilaments (Fig. 3B; Hammond et al. 1986; Lee et al. 1990b; Lee et al. 1991). The postnatal lines are more highly differentiated than the embryonic lines, particularly with respect to the expression of phosphorylated neurofilament isoforms. Of particular interest has been the finding that many of the cell lines that are obtained retain the cholinergic neurotransmitter phenotype typical of cholinergic neurons that are present in the septohippocampal projection, including the expression of the biosynthetic enzyme for acetylcholine, choline acetyltransferase (ChAT; Hammond et al. 1986, 1990). The neuroblastoma line which serves as one of the two parent cells in the fusion has barely detectable ChAT levels. As was seen with cytoskeletal features, the postnatal septal cell lines express higher ChAT levels on the average than do cells obtained from embryonic

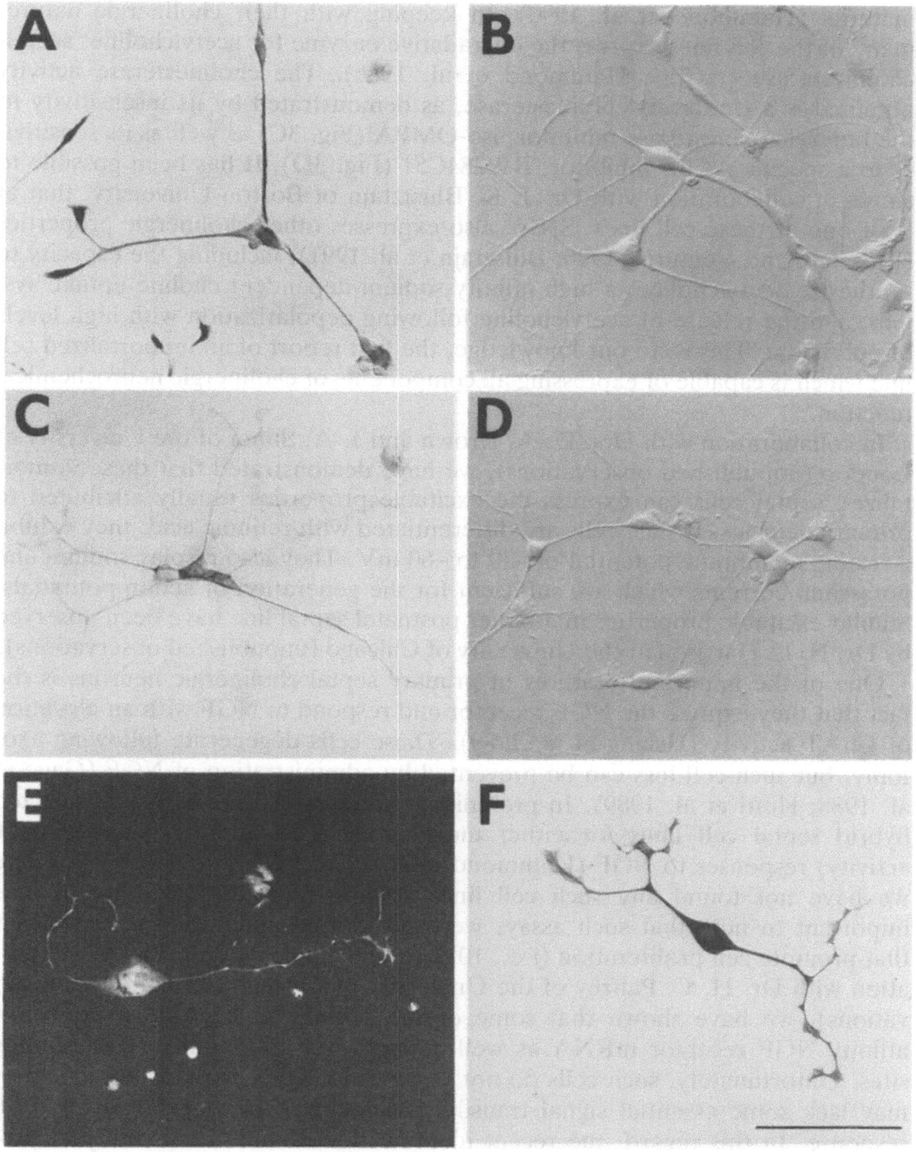

Fig. 3A–F. The neuronal morphological and neurochemical features of the postnatal septal cell line, SN56, are shown in (*A–D*). (A) The presence of neurofilament proteins is demonstrated using immunocytochemical methods and a mixture of antibodies against the low, middle and high molecular weight forms of neurofilament protein. (B) In contrast, there is an absence of immunoreactivity for the astroglial intermediary filament, GFAP. (C) Acetylcholinesterase reaction product is expressed despite treatment with the butyrylcholinesterase inhibitor, ISO-OMPA. (D) Acetylcholinesterase reaction product is not expressed if the cells are treated with the specific acetylcholinesterase inhibitor, BW284C51. (*E*) Endogenous catecholamine histofluorescence is visualized in the embryonic mesencephalic cell line, MN9D, using the Falck-Hillarp histofluorescence technique. (*F*) The catecholamine synthetic enzyme, tyrosine hydroxylase, is visualized in the MN9D cell line using immunocytochemical methods. Calibration bar (A–F): 100 μm

material (Hammond et al. 1990). In keeping with their cholinergic nature, many of the cell lines express the degradative enzyme for acetylcholine, acetylcholinesterase (AChE; Hammond et al. 1986). The cholinesterase activity obtained is a true acetylcholinesterase, as demonstrated by its insensitivity to the butyrylcholinesterase inhibitor, iso-OMPA (Fig. 3C) as well as its sensitivity to a specific AChE inhibitor, BW284C51 (Fig. 3D). It has been possible to show, in collaboration with Dr. J. K. Blustztajn of Boston University, that at least one of these cell lines, SN56, also expresses other cholinergic properties (Blusztajn and Venturini 1990; Blusztajn et al. 1991), including the capacity to synthesize acetylcholine, a high affinity sodium-dependent choline uptake system, and the release of acetylcholine following depolarization with high levels of potassium. This is, to our knowledge, the first report of an immortalized cell line which is capable of expressing all components of cholinergic neurochemical function.

In collaboration with Drs. D. A. Brown and J. A. Simm of the University of London (unpublished observations), we have demonstrated that these immortalized septal cells can express the excitable properties usually attributed to primary neurons. If such cells are differentiated with retinoic acid, they exhibit a resting membrane potential of –30 to –60 mV. They also display sodium and potassium currents which are sufficient for the generation of action potentials. Similar excitable properties in another postnatal septal line have been observed by Dr. N. L. Harrison at the University of Chicago (unpublished observations).

One of the important features of primary septal cholinergic neurons is the fact that they express the NGF receptor and respond to NGF with an elevation of ChAT activity (Hsiang et al. 1989). These cells degenerate following axotomy, but such cell loss can be prevented by administration of NGF (Gage et al. 1988; Hefti et al. 1989). In preliminary studies we have screened selected hybrid septal cell lines for either morphological or neurochemical (ChAT activity) responses to NGF (Hammond et al. 1990; Lee et al. 1990b). To date we have not found any such cell lines responsive to NGF, although it is important to note that such assays were performed under culture conditions that promote cell proliferation (i.e., 10% serum supplementation). In collaboration with Dr. H. C. Palfrey of the University of Chicago (unpublished observations), we have shown that some of the septal cell lines do express low affinity NGF receptor mRNA as well as high and low affinity NGF binding sites. Unfortunately, such cells do not respond to NGF, perhaps because they may lack some essential signal transduction method for elucidating an NGF response. In this regard, the recent identification of the "trk" family of protooncogenes, which are high affinity neurotrophin receptors, has provided further insights into the active receptor complex (Bothwell 1991), and studies are being undertaken to evaluate the septal cell lines for trk expression. Although the septal cell lines currently available do not respond to NGF, they are in fact responsive to other trophic factors. In this regard, Tabira and coworkers have described trophic responses of septal cholinergic neurons to interleukin-3 (IL3; Kamegai et al. 1990b), and these investigators have also found that one of our septal cell lines, SN6.10.2.2, responds to IL3 with an elevation in ChAT activity. In addition, the septal cell line responsive to IL3, like primary septal cells, responds to another cytokine, granulocyte-macrophage colony stimulating factor (GM-CSF; Kamegai et al. 1990a). While the hybrid septal cell lines currently available may not display a full array of responses, the findings to date suggest that such lines will be of use in studying neuronal responses to at least some specific trophic signals.

In addition to the septal cell lines, some 48 hippocampal hybrid cell lines have been generated from embryonic and postnatal brain cells (Lee et al. 1990a). These cell lines exhibit neuronal morphologies and cytoskeletal features as well as the electrophysiological properties typical of primary hippocampal neurons. While hippocampal cell lines have not been extensively studied with respect to neurotransmitter properties, they do elaborate NGF, an essential trophic signal. The postnatal hippocampal lines have, in general, higher levels of NGF protein than do the embryonic lines.

We have also examined the hybrid hippocampal lines for the expression of non-NGF trophic factors. To date, some 22 hippocampal lines as well as the N18TG2 cell line have been screened using a microculture bioassay procedure which permits the examination of conditioned media or subcellular fractions of hippocampal cell lines for the presence of molecules which might elevate ChAT activity in primary septal cholinergic neurons (Lee et al. 1990a; Wainer et al. 1990). Most hippocampal lines show little effect (less than a 75% increase in ChAT activity; however, the conditioned media or crude membrane preparation from one of these lines, HN10, produces a four-fold increase in ChAT activity in this microculture bioassay system. The increase in activity is not neutralized by antibodies against NGF and is not attributable to either epidermal growth factor or basic fibroblast growth factor. The activity is, however, inactivated by proteolysis. The discovery of a number of additional factors, such as brain derived neurotrophic factor (BDNF) and neurotrophin (NT-3), which are members of a family of NGF-related trophic factors, the neurotrophins (Hohn et al. 1990; Roback et al. 1991), raises the possibility that a combination of such factors may be responsible for HN10 activity in raising ChAT levels. For this reason, the available hippocampal lines are being screened for expression of BDNF and NT-3.

Clonal Hybrid Cell Lines of the Nigrostriatal Pathway

Few neuronal projections have elicited as much interest as the dopaminergic inputs to a variety of telencephalic structures of the basal ganglia and neocortex (Fig. 1B). The dopaminergic neurons of the substantia nigra, whose degeneration is apparent in Parkinson's disease (Hornykiewicz 1973), are also believed to subserve a variety of other physiological functions as well as fundamental mental processes whose disruption can result in diseases as extensive as schizophrenia (Hornykiewicz 1978). In contrast to the situation with the cholinergic neurons and their dependence of NGF, little is known regarding the trophic interactions which modulate the establishment and maintenance of the dopaminergic central projections. The recent discovery of the effects of BDNF on dopaminergic neuronal survival in culture (Hyman et al. 1991; Knusel et al.

Table 2. Properties of dopamine-containing mesencephalic hybrid cell lines[a]

Cytoskeleton:	NFP	+
	GFAP	−
Excitability:	Na^+-Currents and action potentials	+
Neurochemical profile:	Dopamine/norepinephrine	+
	Tyrosine hydroxylase	+
	Dopamine Uptake	+
	K^+-stimulated DA release	(not tested)
	MAO	+
	Reserpine and MPP^+ sensitivity	+
Growth factor response:	BDNF	(not tested)
Target vs. non-target cell recognition:		+
Synapse formation:		(not tested)

[a] Abbreviations: NFP, neurofilament proteins; GFAP, glial fibrillary acedic protein; DA, dopamine, MAO, monoamine oxidase; MPP^+, the active neurotoxic metabolite of MPTP, N-methyl-4-phenyl-1,2,3,6-tetrahydropyridine; BDNF, brain derived neurotrophic factor

1991) is the first evidence that the survival of these neurons may depend on some neurotrophic factor.

By use of the cell fusion protocol we have recently succeeded in generating a dopaminergic hybrid cell line (MN9D) from primary mesencephalic cells (Choi et al. 1991b). This cell line exhibits many of the characteristics of primary dopaminergic neurons (Table 2) and should serve as a model for studying the role of chemotrophic signals in the establishment of specific dopaminergic connections as well as a variety of other neuronal functions, including catecholamine storage and synthesis, receptor characterization, and neurotoxicity. This cell line, in contrast to one of its parents, the neuroblastoma line, contains high levels of catecholamines, including dopamine and norepinephreine (Choi et al. 1991b). The levels of catecholamines attained in this line are sufficient to permit the visualization of its cell bodies and processes by standard histofluorescence techniques (Fig. 3B). Similarly, it is possible to visualize tyrosine hydroxylase in such cells by immunocytochemical methods (Fig. 3F). Sufficient DOPA and dopamine are released from these cells so that such compounds are actually detectable in the culture media in which the cells are grown. The MN9D cell line is highly stable and has been found to maintain catecholamine synthesis and release for periods up to a year in continuous passage. The MN9D cell line, like the primary embryonic dopaminergic neuron, is capable of aggregation in rotation-mediated tissue culture (Choi et al. 1991b), and this latter property has made it possible to examine a number of issues of cell-cell interaction between hybrid cell lines and primary cells.

The MN9D cell line has electrophysiological features similar to those of primary dopaminergic neurons. The cells are capable of generating depolarization induced action potentials and voltage-sensitive sodium currents which are very similar to those seen in the intact rodent brain (Choi et al. 1991b).

One feature of the MN9D line is its sensitivity to the depleting action of MPP^+, the active metabolite of N-methyl-4-phenyl-1,2,3,6-tetrahydropyridine (MPTP). This compound, which is a by-product of an attempt to illicitly synthesize meperidine, is a highly toxic substance which produces neurotoxicity to dopaminergic neurons in various species, including humans, mice and monkeys (Choi et al. 1991b; Langston and Irwin 1986). Studies with MPP^+ have also raised the possibility that Parkinson's disease may be in part an environmental or industrial toxicological problem. The fact that the MN9D cell is sensitive to this neurotoxin suggests that such cell lines may provide useful models of monoclonal hybrid cells for studying various mechanisms of neurotoxicity in environmentally induced disease. Preliminary studies reveal that low concentrations of MPP^+ not only decrease catecholamines but may also elicit cell death in the MN9D cell line (Choi et al. 1991a).

With respect to the issue of the use of cell lines as models for gene transfer studies, it is important to determine whether or not specific cell lines can be used to examine trophic interactions between specific neuronal cell types and their targets. If a cell line is to be of use in determining the mechanisms of trophic interactions, it should be capable of responding to signals derived from appropriate target or non-target cells. In recent studies, it has been possible to show that the MN9D cell, which is derived from mesencephalic dopaminergic neurons, is capable of demonstrating a differential response to cells of the corpus striatum or neocortex to which dopaminergic neurons project, as compared to optic tectum or thalamus, anatomic subdivisions of the brain which are not innervated by primary dopaminergic neurons. This response has been demonstrated using a three-dimensional reaggregate system in which primary cells and cell lines can be aggregated together to form small balls of tissue (300 micron in diameter) consisting of a mixture of primary cells and appropriate hybrid cells. In such a system, it is therefore possible to examine the potential interaction between hybrid cells and selected primary cells from various subdivisions of brain. When MN9D cells are coaggregated with cells of corpus striatum, frontal cortex or occipital cortex, all of which are target cells for dopaminergic neurons, they retain their high levels of dopamine and are readily visualized under fluorescence microscopy (Fig. 4; Choi et al. 1990). In contrast, if MN9D cells are coaggregated with non-target cells of the optic tectum or thalamus, they exhibit a marked reduction (approximately 10–20 fold decrease) in dopamine which is so severe that these cells can no longer be visualized under fluorescence microscopy. Preliminary studies (Choi et al., unpublished observations) suggest that the reduction in catecholamine level is secondary to a down-regulation of mRNA for tyrosine hydroxylase, the rate-limiting enzyme in catecholamine biosynthesis. In contrast, PC12 cells, which also synthesize dopamine, do not exhibit any such modulation of neurotransmitter levels when coaggregated with primary cells from these brain regions. The selective effect of dopaminergic non-target cells on the mesencephalic-derived dopaminergic cell line, MN9D, provides direct evidence that immortalized neuronal hybrid cell lines can actually serve as model systems for the examination of putative trophic signals which may be necessary for the elabora-

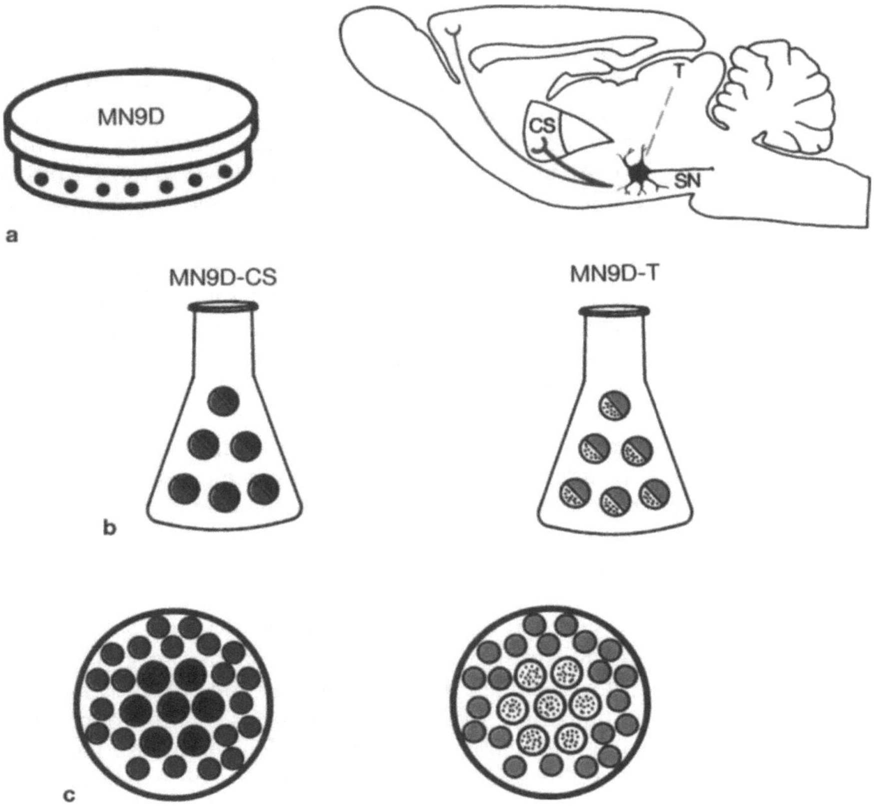

Fig. 4a–c. Schematic illustration of aggregates consisting of dopamine-containing MN9D hybrid cells with appropriate or inappropriate target cells for primary central dopaminergic neurons. (*a*) MN9D hybrid cells (blue) growing in monolayer culture are trypsinized and dissociated into a single cell suspension. The corpus striatum (CS), an appropriate dopaminergic target, and the optic tectum (T), an inappropriate dopaminergic target, are dissected from embryonic brains, trypsinized and dissociated into single cell suspensions. (*b*) Cell suspensions of MN9D (blue) and CS (red) cells, or of MN9D (blue) and T (green) cells, are placed into flasks in rotatory culture, whereupon aggregates (approximately 300 μm) are formed. (*c*) Examination of aggregate sections following processing for Falck-Hillarp histofluorescence showns that MN9D hybrid cells are intensely fluorescent (solid circles) in MN9D-CS aggregates (left section). In contrast, MN9D hybrid cells in MN9D-T aggregates (right section) show no fluorescence or are only weakly fluorescent (stippled blue circles).

tion of the complex and highly selective patterning of axonal connections in the intact brain.

As with the septohippocampal system, it has also been possible to obtain hybrid cell lines from the corpus striatum, the principal target cell population for dopaminergic neurons. In collaboration with Dr. Bruce Perry of the University of Chicago these cells have been shown to express D-1 receptor binding sites which are characteristic of the dopaminergic target cells of the corpus striatum. The binding density of the striatal lines obtained increases by approximately 3–4 fold in the presence of a differentiating agent such as n-butyrate (Wainwright et al. 1990). The parent neuroblastoma line does contain approximately a third as much D-1 receptor binding as the striatal cell lines, but this binding density is not affected by treatment with n-butyrate. The D-1 receptors on the hybrid cells appear to be coupled to adenylate cyclase, since 10^{-4} M dopamine in the presence of 10^{-7} M haloperidol increases adenylate cyclase activity in these cells by some 50% (Wainwright et al. 1990). Given these findings on dopaminergic receptor binding densities and their modulation, the striatal lines obtained to date should be useful for the examination of the developmental regulation of dopaminergic receptor/effector systems as well as serving as a source of enriched cells for the cloning of the mouse dopamine receptor.

Conclusions and Future Directions

We have discussed here the generation of CNS clonal lines using the somatic cell fusion approach. Examples of how this technique has been employed to establish stable cell lines from the septum and hippocampus as well as from the mesencephalon and striatum are provided. The cell fusion technique does not appear to be limited to primary cell populations that are still mitotically active. The generation of postnatal septal cell lines that exhibit all of the biosynthetic machinery necessary for the synthesis and release of ACh strongly suggests that the "immortalization" event involves a primary cell that is already postmitotic and committed to a particular differentiation pathway. The ability to immortalize a neuron at a variety of stages (pre- and postnatal) in development provides an important tool for investigating the cellular and molecular mechanisms that mediate the formation of specific neural circuits. The finding that the dopaminergic cell line, MN9D, can respond to non-target cells and therefore distinguish between target and non-target cells is particularly intriguing with regard to obtaining cell populations that will allow for detailed cellular and molecular studies of neuronal differentiation and cell recognition.

As described here, an array of cellular models has been generated which should be useful in assessing the molecular basis of neuronal survival and the establishment and maintenance of specific central synaptic connections in health and disease. Nevertheless, every methodology and experimental approach has its limitations. In general, it is important to remember that all cell lines, irrespective of the manner in which they were generated, are not composed of primary brain cells, but represent only approximations of such cells. It is therefore essential that any observations made employing cell lines be tested in primary cells in culture and ultimately in the intact brain. Despite this limitation, cell lines represent an unlimited source of material for analyses that cannot normally be performed in either primary culture systems or in the intact brain. One problem with the fusion approach described is that the growth characteristics of the hybrid cell lines are more difficult to control than retrovirally transduced cell lines. This disadvantage undoubtedly is secondary to the overtly malignant nature of the neuroblastoma cell line that was used as one of the fusion partners. Another problem is that the use of a peripheral neural tumor cell line, the N18TG2 neuroblastoma, confers on the hybrid cell line properties that normally do not participate in the development of specific central neuronal pathways. Although the hybrid dopaminergic MN9D cell does exhibit cell-recognition properties, probably due to its derivation from a mesencephalic dopamine neuron, it is not clear at this point whether the influence of the fusion partner has limited the scope of these recognition properties. For example, it has not been possible to demonstrate a positive effect of dopaminergic targets on these cell lines. In this regard, recent work suggests that cell of a particular lineage express regulatory proteins which repress the transcription of genes specific to other lineages (Schafer et al. 1990).

It will be important to develop technologies in the future that will overcome the limitations of the currently available cell lines. New strategies are needed for neuronal cell immortalization beyond those already available. One approach is to utilize the information and technology developed from both retroviral-mediated oncogene transduction and cell fusion techniques, as illustrated schematically in Figure 5. Progenitor cell lines can be established from specific brain regions using the former approach and would allow greater control over the growth characteristics of the resulting cell line. In particular, the use of the temperature-sensitive mutant of the immortalizing SV40 large T antigen could provide a precise handle on cell division (Lendahl and McKay 1990; Cepko 1989). The antigen is stable at 33 °C and promotes cell division, but the antigen is inactivated at 37–39 °C, and then the cells are no longer stimulated to divide. Using a retroviral construct containing this gene, one can develop and characterize a series of progenitor cell lines from a particular brain region of interest (Eves et al. 1990, 1991; Tucker et al. 1990, 1991). As with other cell lines that have been developed with this technique, one would not anticipate that such progenitors of either neuronal or glial origin would differentiate to any great extent, but such cells should retain the lineage features that are unique to the brain region from which they are derived. Is should be possible to select and mutate a particular progenitor cell line of interest so that it can be selected against in culture. Mutations which eliminate the function of the HPRT gene are particularly advantagenous since this gene is X-linked and therefore its inactivation requires only one mutational event. Agents such as ethane methyl sulfonate, a chemical mutagen, are quite efficient in generating HPRT-deficient mutants. HPRT-deficient cells can then be employed as a fusion partner for primary cells from the same brain region, as was done with the somatic fusion products described above, but at a later stage of development. Such an approach would obviate the need for neuroblastomas as a fusion partner and provide cells containing a growth-promoting gene that is better defined and more amenable to control, while allowing for the preservation of region-associated cell lineage. The resulting hybrid cell lines should express a more highly differentiated phenotype because of the potential of fusing the progenitor line to more mature primary neurons.

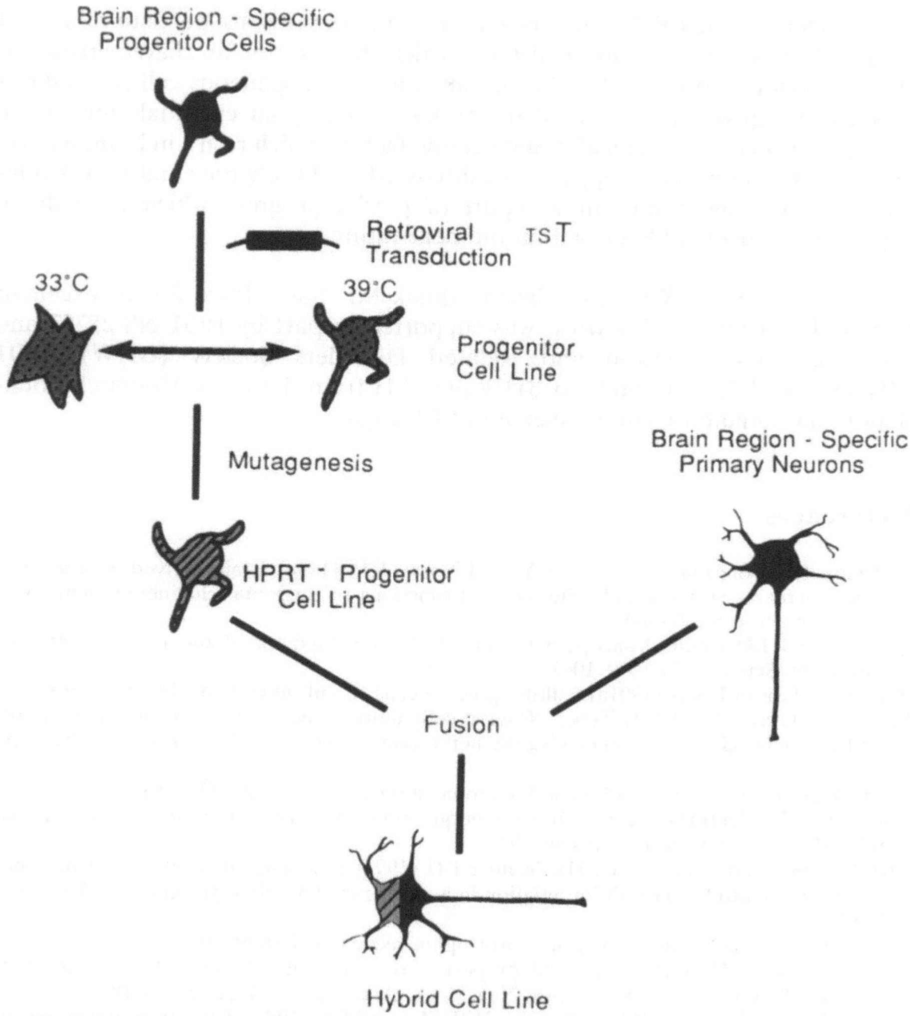

Fig. 5. Schematic illustration of the transduction/fusion strategy for producing CNS clonal cell lines. Primary embryonic brain cells from a specific brain region (black, short processes) are infected with retroviruses containing a temperature-sensitive "immortalizing" oncogene resulting in a "progenitor" cell line (red stippled). The progenitor cell line is treated with a chemical mutagen so that the HPRT gene is functionally eliminated. The HPRT$^-$ progenitor cell line (green hatched) is then fused using polyethylene glycol (PEG) to mature primary neurons (black, long, processes) from the same brain region. Utilizing this oncogene transduction technique, the resulting hybrid cell line (green hatched/black filled) should be more amenable to growth control and should retain the lineage features inherent in the brain region of interest.

Although the feasibility of such a strategy remains to be determined, it would if successful, provide cell lines which display similar characteristics to those of primary brain cells. The production of homogeneous cell populations that can be grown in unlimited quantities is clearly an essential step in the determination of the molecular and genetic factors which result in brain pathology. The techniques and approaches discussed in this chapter and others relevant to this issue represent a report of partial progress which is likely to experience considerable growth in the near future.

Acknowledgements. We would like to thank Mr. Steve Price for his excellent technical assistance. This work was supported in part by RO1 NS-25787 and the Alzheimer's Disease and Related Disorders Society (BHW), RO1 MH28942 (AH), and grants to BHW and AH from the Brain Research Foundation, an affiliate of The University of Chicago.

References

Alderson RF, Alterman AL, Barde Y-A, Lindsay RM (1990) Brain-derived neurotrophic factor increases survival and differentiated functions of rat septal cholinergic neurons in culture. Neuron 5:297–306

Alt FK, GD Blackwell, Yancopoulos GD (1987) Development of the primary antibody repertoire. Science 238:1079–1087

Banker G, Goslin K eds (1991a) Culturing nerve cells. Cambridge, MA, The MIT Press

Banker G, Goslin K (1991b) Types of nerve cell cultures, their advantages and limitations. In: Banker G, Goslin K (eds) Culturing nerve cells. Cambridge, MA, The MIT Press, pp 11–39

Barde Y-A (1989) Trophic factors and neuronal survival. Neuron 2:1525–1534

Blusztajn JK, Venturini A (1990) Cholinergic properties of a murine septal cell line SN56.B5.G4. Soc Neurosci Abs 16:199

Blusztajn JK, Venturini A, Lee HJ, Wainer BH (1991) Independent regulation of morphological and neurochemical differentiation in a murine cholinergic septal cell line. FASEB J 5:A456.

Bothwell M (1991) Keeping track of neurotrophin receptors. Cell 65:915–918

Bottenstein JE (1981) Differentiated properties of neuronal cell lines. In: G Sato (ed) Functionally differentiated cell lines. New York, Alan R, Liss, Inc, pp 155–184

Carden MJ, Trojanowski JQ, Schlaepfer WW, Lee VM-Y (1987) Two-stage expression of neurofilament polypeptides during rat neurogenesis with early establishment of adult phosphorylation patterns. J Neurosci 7:3489–3504

Cepko CL (1989) Immortalization of neural cells via retroviral-mediated oncogene transduction. Ann Rev Neurosci 12:47–65

Choi HK, Won LA, Heller A (1990) The dopamine content of immortalized hybrid neurons is dependent on the presence of target tissue. Soc Neurosci Abs. 16:996, 412.20

Choi HK, Won L, Hoffmann PC, Heller A (1991a) Sensitivity of an immortalized dopaminergic cell line to MPP^{+} cytotoxicity. Soc Neurosci Abs 17:1275

Choi HK, Won LA, Kontur PJ, Hammond DN, Fox AP, Wainer BH, Hoffmann PC, Heller A (1991b) Immortalization of embryonic mesencephalic dopaminergic neurons by somatic cell fusion. Brain Res 552:67–76

Cotman CW, Matthews DA, Taylor D, Lynch G (1973) Synaptic rearrangement in the dentate gyrus: histochemical evidence of adjustments after lesions in immature and adult rats. Proc Natl Acad Sci USA 70:3473–3477

Decker MW, McGaugh JL (1991) The role of interactions between the cholinergic system and other neuromodulatory systems in learning and memory. Synapse 7:151–168

Delacour J, Houcine O, Costa JC (1990) Evidence for a cholinergic mechanism of "learned" changes in the responses of barrel field neurons of the awake and undrugged rat. Neuroscience 34:1–8

Eves EM, Lee HJ, Tucker MS, Rosner MR, Wainer BH (1990) Immortal cell lines from embryonic rat hippocampal precursor cells. Soc Neurosci Abs. 16:1149

Eves EM, Kwon J, Wainer BH (1991) Induction of rapid differentiation in immortalized rat septal cell lines. Soc Neurosci Abs 17:37

Fischer W, Wictorin K, Bjorklund A, Williams LR, Varon S, Gage FH (19897) Amelioration of cholinergic neuron atrophy and spatial memory impairment in aged rats by nerve growth factor. Nature 329:65–68

Gage FH, Armstrong M, Williams LR, Varon S (1988) Morphological response of axotomized septal neurons to nerve growth factor. J Comp Neurol 269:147–155

Gage FH, Kawaja MD, Fisher LJ (1991) Genetically modified cells: applications for intracerebral grafting. Trends Neurosci 14:328–333

Greene LA, Shain W, Chalazonitis X, Breakfield J, Minna HG, Coon M, Nirenberg M (1975) Neuronal properties of hybrid neuroblastoma X sympathetic ganglion cells. Proc Natl Acad Sci USA 72:4923–4927

Greene LA, Sobeih MM, Teng KK (1991) Methologies for the culture and experimental use of the PC12 rat pheochromocytoma cell line. In: Banker G, Goslin K (eds) Culturing nerve cells. Cambridge, MA, The MIT Press, pp 207–226

Hammond DN, Wainer BH, Tonsgard JH, Heller A (1986) Neuronal properties of clonal hybrid cell lines derived from central cholinergic neurons. Science 234:1237–1240

Hammond DN, Lee HJ, Tonsgard JH, Wainer BH (1990) Development and characterization of clonal cell lines from septal cholinergic neurons. Brain Res 512:190–200

Hefti F, Hartikka J, Knusel B (1989) Function of neurotrophic factors in the adult and aging brain and their possible use in the treatment of neurodegenerative disease. Neurobiol Aging 10:515–533

Hohn A, Leibrock J, Bailey K, Barde Y-A (1990) Identification and characterization of a novel member of the nerve growth factor/brain-derived neurotrophic factor family. Nature 344:339–341

Hornykiewicz, O (1973) Parkinson's disease: from brain homogenate to treatment. Fed Proc 32:183–190

Hornykiewicz O (1978) Psychpharmacological implications of dopamine and dopamine antagonists: A critical evaluation of current evidence. Neuroscience 3:773–783

Hsiang J, Heller A, Hoffmann PC, Mobley WC, Wainer BH (1989) The effects of nerve growth factor on the development of septal cholinergic neurons in reaggregate cell cultures. Neuroscience 29:209–223

Hyman C, Hofer M, Barde Y-A, Juhasz M, Yancopoulos GD, Squinto SP, Lindsay RM (1991) BDNF is a neurotrophic factor for dopaminergic neurons of the substantia nigra. Nature 350:230–232

Kamegai M, Konishi Y, Tabira T (1990a) Trophic effect of granulocyte-macrophage colony-stimulating factor on central cholinergic neurons in vitro. Brain Res 532:323–325

Kamegai M, Niijima K, Kunishita T, Nishizawa M, Ogawa M, Araki M, Ueki Y, Konishi Y, Tabira T (1990b) Interleukin 3 as a trophic factor for central cholinergic neurons in vitro and in vivo. Neuron 4:429–436

Katzman R, Saitoh T (1991) Advances in Alzheimer's Disease. FASEB J 5 (3):278–286

Kesslak JP, Frederickson CJ, Gage F-H (1987) Quantification of hippocampal noradrenaline and zinc changes after selective cell destruction. Exp Brain Res 67:77–84

Knusel B, Winslow JW, Rosenthal JW, Burton LE, Seid DP, Nikolics K, Hefti F (1991) Promotion of central cholinergic and dopaminergic neuron differentiation by brain-derived neurotrophic factor but not neurotrophin 3. Proc Natl Acad Sci USA 88:961–965

Kohler G, Milstein C (1975) Continuous cultures of fused cells secreting antibody of defined specificity. Nature 256:495–497

Langston JW, Irwin J (1986) MPTP. Current concepts and controversies. Clin Neuropharmacol 9:485–507

Lee HJ, Hammond DN, Large TH, Sim JA, Brown DA, Otten UH, Wainer BH (1990a) Neuronal properties and trophic activities of immortalized hippocampal cells from embryonic and young adult mice. J Neurosci 10:1779–1787

Lee HJ, Hammond DN, Large TH, Wainer BH (1990b) Immortalized young adult neurons from the medial septal region: Production and characterization. Dev Brain Res 52:219–228

Lee HJ, Elliot GJ, Hammond DN, Lee VM-Y, Wainer BH (1991) Constitutive expression of the mature array of neurofilament by a CNS neuronal cell line. Brain Res 1–12

Lendahl U, McKay RDG (1990) The use of cell lines in neurobiology. TINS 13:132–137

Lindvall O (1991) Prospects of transplantation in human neurodegenerative diseases. TINS 14:376–384

Maisonpierre P, Belluscio L, Friedman B, Alderson RF, Wiegand SJ, Furth ME, Lindsay RM, Yancopoulos GD (199O) NT-3, BDNF, and NGF in the developing rat nervous system: Parallel as well as reciprocal patterns of expression. Neuron 5:501–509

Marx J (1990) NGF and Alzheimer's: Hopes and Fears. Science 247:408–410

Nilsson OG, Shapiro ML, Gage FH, Olton DS, Bjorklund A (1987) Spatial learning and memory following fimbria-fornix transection and grafting of fetal septal neurons to the hippocampus. Exp Brain Res 67:195–215

Paige CJ, Wu GE (1989) The B cell repertoire. FASEB J 3:1818–1824

Phelps CH, Gage FH, Growdon JH, Hefti F, Harbaugh R, Johnston MV, Khachaturian ZS, Mobley WC, Price DL, Raskind M, Simpkins J, Thal LJ, Woodcock J (1989) Potential use of nerve growth factor treat Alzheimer's disease. Neurobiol Aging 10:205–207

Platika D, Boulos MH, Baizer L, Fishmann MC (1985) Neuronal traits of clonal cell lines derived by fusion of dorsal root ganglia neurons with neuroblastoma cells. Proc Natl Acad Sci USA 82:3499–3503

Pletscher A (1990) Levodopa treatment of Parkinson's syndrome: past and future. Adv Neurol 53:469–473

Purves D (1988) Body and brain: A trophic theory of neural connections. Cambridge, Massachusetts: Harvard University Press

Roback JH, Palfrey HC, Wainer BH (1991) Nerve growth factor and the neurotrophin family: Evolving roles in the central nervous system. Curr Topics Devel Biol, in press

Ronnett GV, Hester LD, Nye JS, Connors K, Snyder SH (1990) Human cortical neuronal cell line: Establishment from a patient with unilateral megalencephaly. Science 248:603–605

Rosenberg MB, Friedman T, Robertson RC, Tuszynski M, Wolff JA, Breakefield XO, Gage FH (1988) Grafting genetically modified cells to the damaged brain: restorative effects of NGF expression. Science 242:1575–1578

Rye DB, Wainer BH, Mesulam M-M, Mufson EJ, Saper CB (1984) Cortical projections arising from the basal forebrain: a study of cholinergic and noncholinergic components employing combined retrograde tracing and immunohistochemical localization of choline acetyltransferase. Neuroscience 13:627–643

Saper CB, Wainer BH, German DC (1987) Axonal and transneuronal transport in the transmission of neurological disease: potential role in system degenerations, including Alzheimer's disease. Neursocience 23:389–398

Schafer BW, Blakely BT, Darlington GJ, Blau HM (1990) Effect of cell history on response to helix-loop-helix family of myogenic regulators. Nature 344:454–458

Schubert D (1984) Introduction. In: Schubert D (ed) Development biology of cultured nerve, muscle and glia. New York, John Wiley & Sons, pp 1–25

Shay JW (1982) Techniques in somatic cell genetics. New York, Plenum Press

Tishler AS, Greene LA (1975) Nerve growth factor-induced process formation by cultured rat pheochromocytoma cells. Nature 258:341–342

Tucker MS, Eves EM, Hou XY, Wainer BH, Rosner MR (1990) Expression of epidermal growth factor receptor in immortalized rat hippocampal cell lines. Soc Neurosci Abs 16:1149

Tucker MS, Eves EM, Hou XY, Wainer BH, Rosner MR (1991) Expression and regulation of epidermal growth factor receptor in differentiating rat hippocampal cell lines. FASEB J 5:A1622

Wainer BH, Levey AI, Rye DB, Mesulam M-M, Mufson EJ (1985) Cholinergic and non-cholinergic septohippocampal pathways. Neurosci Lett 54:45–52

Wainer BH, Lee HJ, Roback JD, Hammond DN (1990) In vitro cell cultures as a model of the basal forebrain. In: Kalivas PW, Napier TC (eds) The basal forebrain: anatomy to function. New York, Plenum Press

Wainer BH, Mesulam M-M (1990) Ascending cholinergic pathways in the rat brain. In: Steriade M, Biesold D (eds) Brain cholinergic systems. New York, Oxford University Press

Wainwright MW, Perry BD, Kontur P, Heller A (1990) Expression of D_1-dopamine receptor binding sites in an immortalized murine corpus striatum cell line. Soc Neurosci Abs 16:646,272.11

Resistance of SOD-Transgenic Mice to Oxidative Stress

C. J. Epstein, P. H. Chan, J. L. Cadet, E. Carlson, S. Chen, L. Chu, S. Fahn, V. Jackson-Lewis, H. Kinouchi, V. Kostic, K. Kujirai, A. Mizui, A. Naini, S. Przedborski and *G. Yang*

Summary

The activity of CuZn-superoxide dismutase (CuZnSOD) in the brain and other tissues of mice has been increased between 1.5- and 5-fold by the insertion of one or more human CuZnSOD transgenes into the mouse genome. Although physically normal, these transgenic mice exhibit several differences from control mice when exposed to known or presumed forms of oxidative stress. After a 30-second cold injury to the cerebral cortex, transgenic mice with approximately 3-times normal CuZnSOD activity have less cerebral edema, a smaller increase in vascular permeability, and a smaller area of infarction than do the control animals. These transgenic animals are also protected to a significant degree against the effects of transient focal cerebral ischemia in the middle cerebral artery distribution, with sparing particularly of the penumbra region surrounding the ischemic brain tissue. Furthermore, in contrast to control animals, the injection of 1-methyl-4-phenyl-1,2,3,6-tetrahydropyridine (MPTP) into transgenic mice does not induce any depletion in striatal levels of dopamine and its metabolites or decreases in dopamine uptake sites. MPTP is an agent which causes changes in the dopaminergic system similar to those observed in Parkinson's disease. Cultured primary cortical neurons from transgenic fetuses demonstrate a significant reduction in the toxic effects of exposure to 0.5 mM glutamate.

In addition to the protection afforded against the several types of acute stress just enumerated, increased CuZnSOD activity in the brain may also have long-term effects, the functional consequences of which are unknown. Such effects have been detected with regard to the numbers of neurotensin binding sites in the brains of aged transgenic animals. In certain regions, the transgenic animals display age-dependent reductions in binding which were not observed in control animals, whereas in other regions the transgenic animals did not have the reduction that the control animals had. Similar results, although involving other regions, were also obtained for the binding of μ opioid receptors.

Introduction

Oxygen free radicals have been implicated in many pathological processes affecting the brain and other organs and tissues, but it has not always been possible to examine their roles directly. One approach to such an examiniation is to perturb one or more of the components of the metabolic pathway that detoxifies such free radicals. Transgenic animal technology has made it possible to constitutively alter the level of the first enzyme in the detoxification of superoxide anions, superoxide dismutase (SOD), and to examine the effects of such an alteration on the response to agents, both physical and chemical, which are thought to produce deleterious effects through the mediation of oxygen radicals.

The superoxide dismutases catalyze the dismutation of superoxide radicals, O_2^-, to H_2O_2, which is then metabolized to water by glutathione peroxidase or catalase. They are generally regarded as protective enzymes, and numerous examples exist of beneficial effects of exogenously added enzyme (summarized in Halliwell and Gutteridge 1989). However, because the H_2O_2 generated as a result of SOD activity is itself toxic and may, in combination with O_2^-, give rise to the even more dangerous and reactive hydroxyl radical (OH·), it has been argued that increased SOD may, under certain circumstances, have a deleterious effect (Sinet et al. 1979; Sinet 1982). In humans, the question remains whether increased SOD activity has any effect on the balance of oxygen metabolites and, if so, what these effects may be. Animals models are now being used to investigate this problem.

For the studies to be reported below, we used transgenic mice derived from the founder mouse, TgHS/SF-218, described by Epstein et al. (1987). The human CuZnSOD transgene (in eight copies) was carried on a CD-1 background, and CuZnSOD activities were increased approximately 2.5 to 3-fold in the nervous system (as well in fibroblasts and erythrocytes) (Chan et al. 1991; Przedborski et al. 1992; Huang et al. 1992).

Oxidative Stress Induced by Physical Agents In Vivo

Two experimental paradigms in which oxygen free radicals are presumed to play an important role have been employed to produce brain injury. The first was a cold injury trauma model in which a 0.5 cm concave probe cooled to –50 °C was applied directly to the calvarium for 30 seconds. This is regarded as a highly reproducible method for producing vasogenic brain edema. When analyzed two hours after the injury, transgenic brains exhibited 60% less edema (as measured by water content) and 47.3% less increase in vascular permeability (as measured by extravascation of injected Evans blue dye) than did non-transgenic controls (Table 1). Similarly, the area of the resulting cortical infarct 24 hours after injury was decreased by 51.5% (Chan et al. 1991).

Quite comparable results were obtained with the second model system, a stroke model based on middle cerebral artery ischemia and reperfusion (Kinouchi et al. 1991). In this approach, the left middle cerebral artery is occluded by electrocoagulation, the left common carotid is ligated, and the right common carotid artery is occluded for one hour with a micro-aneurysmal clip. This method results in infarction of the brain in the distribution of the left middle

Table 1. Reduction of brain edema, vascular permeability, and infarct size in transgenic mice subjected to cold injury[a]

	Non-transgenic		Transgenic		Reduction in transgenic mice
	Ipsilateral	Contra-lateral	Ipsilateral	Contra-lateral	
Water content, % (2 hours)	80.0	77.5	78.5	77.5	
Difference	2.5		1.0		60.0%
Evans blue content, µg/g (2 hours)	130.7	23.6	87.1	30.7	
Difference	107.1		56.4		47.3%
Infarct area, % (24 hours)	6.6		3.2		51.5%

[a] Data from Chan et al. (1991).

Table 2. Reduction of brain edema and infarct size in transgenic mice subjected to ischemia and reperfusion[a]

		Non-transgenic		Transgenic		Reduction in transgenic mice
		Ipsilateral	Contra-lateral	Ipsilateral	Contra-lateral	
Water content, % (24 hours)	MCA	85.4	78.3	82.9	78.2	
Difference		7.1		4.7		33.8%
	ACA	81.9	78.3	80.3	78.2	
Difference		3.6		2.1		41.7%
Infarct volume, % (24 hours)		18.4		13.7		25.5%

[a] Abbreviations: ACA, anterior cerebral artery territory; MCA, middle cerebral artery territory. Data from Kinouchi et al. (1991)

cerebral artery (MCA) with extension into the region of the left anterior cerebral artery (ACA) distribution. When analyzed 24 hours after injury, the amount of brain edema in the transgenic brains was decreased 33.8% in the MCA territory and 41.7% in ACA territory (Table 2). Infarct volume, expressed as the proportion of total hemispheric volume, was decreased 25.5%. The decrease in the volume of the infarct is believed to result from partial protection of the so-called "penumbra" zone, which surrounds the region of complete cell death and in which the degree of ischemia is less severe. In non-transgenic animals, this zone is initially viable, but the neurons and glia die with time. Although the levels of the antioxidants, reduced glutathione and reduced ascorbate, were decreased in the ischemic hemisphere, the drop in the surrounding cortex and striatum was significantly less in the transgenic than in the non-transgenic mice. This finding suggests that the increased CuZnSOD activity, by metabolizing superoxide radicals, reduces the oxidative challenge to these compounds and, by inference, to the brain itself.

The results of these experiments are quite consistent with a large body of literature implicating oxygen free radicals in the pathogenesis of the deleterious consequences of brain trauma or ischemia (Kontos 1985; Hall and Braughler 1989; Chan 1989; Choi 1990). However, the present approach differs from approaches used in earlier studies to alter CuZnSOD activity in that the increase in enzyme activity is intracellular and genetically determined, rather than being the result of less well-controlled exogenous administration of CuZn-SOD in liposomes or conjugated to polyethylene glycol (Chan et al. 1987; Liu et al. 1989; Imaizumi et al. 1990).

Neuronal Killing by MPTP

Administration of 1-methyl-4-phenyl-1,2,3,6-tetrahydropyridine (MPTP) to humans and mice has been observed to cause damage to the nigrostriatal dopaminergic system similar to what has been observed in Parkinson disease. The dopaminergic system is vulnerable to this damage because it contains the enzyme, monoamine oxidase-B, which converts MPTP to its active form, MPP^+ (1-methyl-4-phenyl pyridium), and because MPP^+ is actively taken up by dopamine uptake sites (Markey et al. 1984; Javitch et al. 1985). Superoxide radicals have been implicated in the causation of the MPTP-induced damage, possibly being produced in response to inhibition of mitochondrial respiratory complex I (Krishnamoorthy and Hinkle 1988) or by a redox reaction between MPP^+ and another MPTP metabolite, $MPDP^+$ (1-methyl-4-phenyl-2,3-dihyro-pyridium) (Rossetti et al. 1988; Jenner 1989).

Three intraperitoneal injections of 30 mg/kg MPTP-HCl at 24-hour intervals to albino (white) non-transgenic animals resulted, five days after the last injection, in marked decreases in the concentrations of striatal dopamine (decreased 62%) and of the dopamine metabolites, dihydroxyphenylacetic acid (decreased 54.5%) and homovanillic acid (decreased 50%) (Przedborski et al. 1992). Furthermore, the binding of [^{3}H]mazindol, a measure of dopamine binding site, was decreased by 36% in the striatum and 21% in the substantia nigra pars compacta. None of these changes was observed in the brains of transgenic mice, even though both the levels of MPTP and MPP^+ in the brain and the kinetics of MPP^+ uptake by striatal synaptosomes were the same in transgenic and non-transgenic mice (Przedborski et al. 1991; 1992). It appears, therefore, that superoxide radicals are in fact involved in the pathogenesis of the toxic effects of MPTP on dopaminergic cells.

The Behavior of Transgenic Cells In Vitro

Another approach to assessing the effects of increased CuZnSOD activity on the response to oxidative stress is to use cells cultured from transgenic and non-transgenic animals. When primary cortical neurons derived from 15-day fetuses and grown 10–12 days in culture were exposed to 0.5 mM glutamate for five minutes, the transgenic cells were partially protected against the resulting toxic effects. Cell death, as measured by trypan blue exclusion, and cell damage, as measured by release of lactate dehydrogenase, were decreased by about 30% when assayed four hours after exposure (Chan et al. 1990a). Neuronal swelling, as represented by 3-O-methylglucose space, was decreased by 36%. However, these protective effects were somewhat less pronounced when the cultures were examined 24 hours after exposure. At that time, 95% of the non-transgenic and 80% of the transgenic cells had died (Chan et al. 1990b).

The protective effects of the increased CuZnSOD activity in another type of cell derived from transgenic animals has also been demonstrated. In this instance, fibroblasts were cultured from heterozygous transgenic fetuses carrying one transgene (actually, one set of tandemly integrated transgenes), as did all of the animals and cells discussed to this point, as well as from fetuses carrying two of the same transgene (resulting in higher CuZnSOD activity) or carrying other transgenes with the same or lower enzyme activity. When these several sets of fibroblasts were challenged with paraquat, a powerful inducer of superoxide radicals, the degree of protection against the resulting growth inhibition increased with log CuZnSOD activity (Huang et al. 1992). Again, under the conditions of these experiments, greater CuZnSOD activity confers greater resistance to oxidative stress under acute circumstances.

Long-Term Effects of Increased CuZnSOD Activity in the Brain

Although all of the data presented above point to a beneficial effect of increased CuZnSOD activity under conditions of acute oxidative stress, it is also important to look at the long-term consequences of alterations in the balances of enzymes involved in oxygen free radical metabolism. This is so because of the possibility that increased CuZnSOD activity could, by permitting hydrogen peroxide to be produced more rapidly than it normally would, lead to enhancement of lipid peroxidation and other oxidative reactions which would be injurious. Several years ago it was suggested that such might be the case in the pathogenesis of the mental retardation associated with Down syndrome (trisomy 21), in which there is a 1.5-fold increase in the activity of CuZnSOD in all tissues (Sinet et al. 1979; Sinet 1982). Therefore, the brains of young (six weeks) and old (20–21 months) transgenic and non-transgenic mice were analyzed by autoradiographic receptor analysis techniques with regard to the binding of two ligands, neurotensin (NT) and Tyr-D-Ala-Gly-NMe-Phe-Gly-ol (DAGO; Cadet et al. submitted for publication; Kujirai et al. 1991; submitted for publication). NT was chosen because its receptors are located on dopaminergic cell bodies and terminals in some regions of the brain. Since dopamine may produce free radicals either by autooxidation or by metabolism by monoamine oxidase, it was thought that NT receptors located on dopaminergic cell bodies or terminals might be especially vulnerable to damage by oxygen free radicals. Similarly, since a significant proportion of μ opioid receptors, to which DAGO binds, is located in close apposition to dopaminergic terminals and cell bodies in the caudate-putamen, and because catecholamines are considered to cause toxicity by free radical-dependent mechanisms (Cadet et al. 1989; Spina and Cohen 1989), it was thought that these opioid receptors might also be vulnerable. In both instances, it was found that there are significant differences between aged transgenic and non-transgenic animals and, in the case of the μ opioid receptor, between young transgenic and non-transgenic animals as well.

When [^{3}H]NT was used as the ligand, there were no differences between young transgenic and non-transgenic animals in the numbers of binding sites in most regions of the brain (Cadet et al., submitted for publication). Furthermore, for both groups of animals there were parallel age-related decreases in binding sites in the majority of the areas of the brain examined and an age-related increase in the CA2 region of the hippocampus. However, significant age-related decreases in NT binding were observed in the septum, the diagonal band of Broca, and in some divisions of the caudate putamen only in transgenic brains, and in the core area of the nucleus accumbens and the dorsal aspect of the dentate gyrus of the hippocampus only in non-transgenic brains (Fig. 1 and Table 3).

Table 3. Densities of [^{3}H] neurotensin binding sites (in fmol/mg tissue) in regions of the brain showing significant differences among young (6 weeks) and old (21 months) transgenic and non-transgenic mica[a]

	Transgenic		Non-transgenic	
	Young	Old	Young	Old
Basal forebrain				
Lateral septum	161.3	130.0[b]	152.6	150.1
Medial septum	57.9	33.2[b]	44.4	38.9
Diagonal band of Broca	62.6	38.6[b]	53.1	45.5
Candate-putamen				
Anterior	42.7	30.9[b]	43.1	35.7
Middle	45.9	27.9[b]	40.9	35.4
Posterior	37.3	24.1[b]	35.5	28.3
Nucleus accumbens core	33.7	36.9	38.9	20.4[b]
Hippocampus				
Dentate gyrus, distal	18.9	19.1	26.0	7.7[b]

[a] Data from Cadet et al. submitted for publication
[b] Mean for old animals significantly less than for young animals of same type.

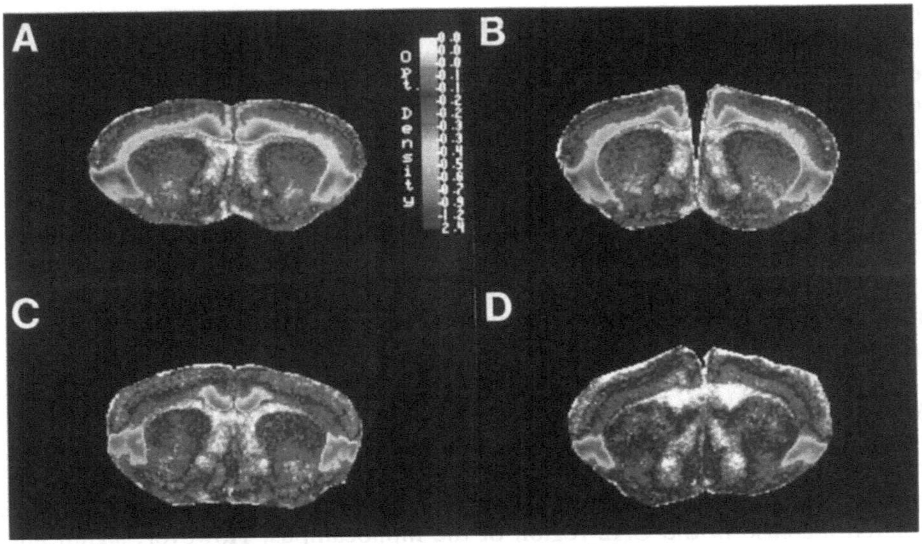

Fig. 1 A–D. Color-coded transforms of autoradiograms of [^{3}H] neurotensin binding at rostral levels of mouse brain. The sections were from non-transgenic (A and C) and SOD-transgenic (B and D) mice at 6 weeks (A and B) and 21 months (C and D) of age. Differences in the laminar distrubution of the NT receptor in the cortex and a marked reduction of NT binding sites with aging are apparent. A summary of the significant differences related to the presence of the SOD-transgene is given in Table 3.

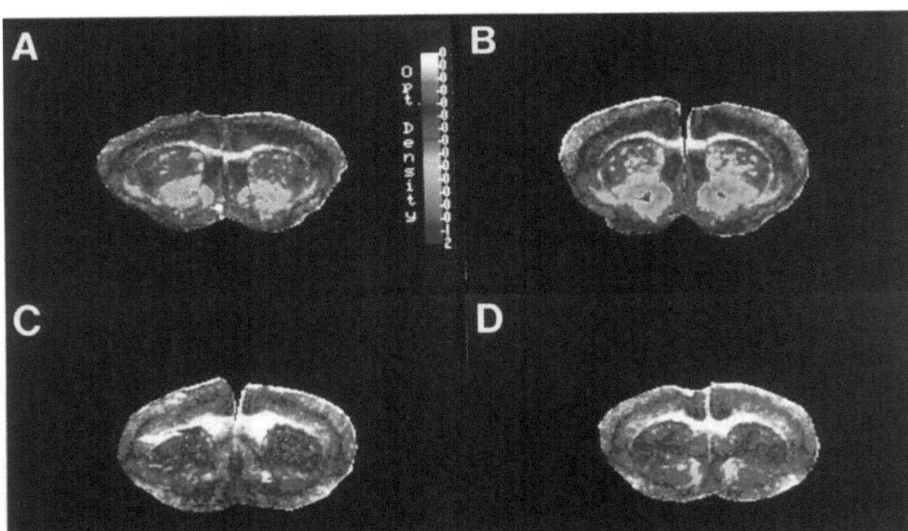

Fig. 2 A–D. Color-coded transforms of autoradiograms of [^{3}H] DAGO binding to mouse brain. The sections were from non-transgenic (A and C) and SOD-transgenic (B and D) mice at 6 weeks (A and B) and 21 months (C and D) of age. A high concentration of μ opioid receptors is present in both the nucleus accumbens core and shell. Significant differences between young and old SOD-transgenic mice in the nucleus accumbens shell are visible, as are general decreases in [^{3}H]DAGO binding with aging. A summary of significant differences related to the presence of the SOD-transgene is given in Table 4.

A similarly complex pattern of findings was obtained when [^{3}H]DAGO was used as the ligand (Fig. 2 and Table 4; Kujirai et al. 1991; submitted for publication. In general, the qualitative distribution of μ opioid receptors was virtually identical in the brains of transgenic and non-transgenic mice. However, young transgenic mice had higher binding densities than did non-transgenic animals in several regions of the brain, including the shell subdivision of the nucleus accumbens, the pars compacta of the substantia nigra, the ventral tegmental area, and the ventral part of the central gray matter. Furthermore, there were more precipitous age-related losses of binding in the latter three areas in transgenic than in non-transgenic brains. By contrast, the decline with age in binding to the shell division of the nucleus accumbens was greater in the non-transgenic brains, and the same was also true of the declines in binding to the core of the nucleus accumbens, the medial septal nucleus, the diagonal band a Broca, and the periclaustrum. Both groups of mice demonstrated parallel decreases in binding in several other regions of the brain.

Although it is possible to speculate on how an elevation in CuZnSOD activity might affect free radical metabolism in various tissues, the precise mechanisms by which the changes just described occur during early development (resulting in altered densities of the μ opioid receptor sites in several regions of the brain) and with aging are not known, and neither are the functional consequences of the alterations described. However, the important

Table 4. Densities of [^{3}H] DAGO binding sites (in fmol/mg tissue) in regions of the brain showing differences among young (6 weeks) and old (21 months) transgenic and non-transgenic mice[a]

	Transgenic		Non-transgenic	
	Young	Old	Young	Old
Nucleus accumbens				
Shell	266.7[b]	229.7[b]	204.0	140.7[c]
Core	276.4	210.7[b, c]	260.9	168.8[c]
Periclaustrum	137.6	139.8[b]	145.9	110.9[c]
Medial septal nucleus	120.9	85.6[b, c]	117.0	58.3[c]
Diagonal band of Broca				
Vertial limb	120.4	90.3[b, c]	115.8	56.8[c]
Horizontal limb	129.4	117.8[b]	136.8	62.9[c]
Substantia nigra pars compactas	73.6[b]	42.7[c]	53.3	41.8
Ventral tegmental area	103.9[b]	65.2[c]	58.9	46.8
Central gray, ventral	120.4[b]	91.6[c]	88.1	76.9

[a] Date from Kujirai et al., submitted for publication
[b] Mean for transgenic animals significantly greater than for non-transgenic of same age
[c] Mean for old animals significantly less than for young animals of same type

point is that long exposure of the brain to increased actitivty of CuZnSOD may lead to changes which are potentially deleterious and may affect the function of the nervous system. Further evidence for this point has been obtained by Ceballos-Picot et al. (1991), who have made a strain of human CuZnSOD-transgenic mouse which has a 1.93-fold increase in the activity of brain CuZn-SOD. They observed a modest (13%) but still significant increase in thiobarbituric acid-reactive material (malondialdehyde), presumed to represent products of lipid peroxidation, in the brains of still young (50–70 days old) transgenic mice.

Discussion

The generation of transgenic animals is certainly not the same as gene therapy, as the latter is presently envisioned, either in intent or in the precise methods generally employed. Nevertheless, both approaches do result in the permanent genetic alteration of host cells with the addition of a new gene or genes to either a normal or abnormal host genome. For those situations in which animal models for human genetic disorders already exist, as in the cases of mice with β-thalassemia and mucopolysaccharidosis VII (β-glucuronidase deficiency), transgenic approaches predict what might be expected if gene therapy could be successfully applied (Costatini et al. 1986; Kyle et al. 1990).

The work described in this report is somewhat different in nature in that the question is not one of replacing a gene that is missing but of increasing the number of copies of a gene already present in the genome. Whether one would actually wish to do so with *SOD1*, the gene for CuZnSOD, is a matter of conjecture, but there has certainly been considerable interest in whether CuZn-SOD could be used therapeutically to decrease the morbidity and mortality associated with a number of conditions in which oxygen free radicals produced by ischemia and reperfusion, as well as by other forms of physical and metabolic perturbation, may be involved (see, for example, Cohen 1989; Jackson et al. 1990; Koke et al. 1990; Daube et al. 1991; Wang et al. 1991; Droy-Lefaix et al. 1991; Das et al. 1991).

Our results indicate that it is not as yet possible to give an unequivocal answer to whether increasing CuZnSOD activity in the brain by genetic means is, on balance, beneficial or deleterious. Under the conditions of our experiments, it appears that a 3-fold increase in CuZnSOD activity in the brain of the mouse has significant protective effects against acute ischemia-reperfusion and trauma-induced damage and against an acute chemical challenge such as is imposed by administration of MPTP. Whether even greater levels of CuZn-SOD would be more or less protective is currently under investigation. The situation is less clear in so far as the long-term effects of increased CuZnSOD activity in the brain are concerned. If the changes observed in the binding of ligands to receptors located on neurons are indicative of functionally deleterious alterations, these findings, coupled with the observations of enhanced membrane lipid peroxidation, would suggest that having more CuZnSOD in the brain might not be entirely beneficial. Further evidence that this may be the case is provided by the analysis of another series of CuZnSOD-transgenic mice prepared by Groner and his collaborators with the same construct used to make the transgenic mice discussed in this report (Epstein et al. 1987). These investigators have found that neuromuscular junctions in the tongues and hind limbs of their transgenic animals are abnormal (Groner et al. 1990; Avraham et al. 1988, 1991) and that the uptake of serotonin by transgenic platelets is diminished (Groner et al. 1990). However, with regard to the possible relevance of the latter observation to the central nervous system, it should be noted that the content of serotonin in the striatum of our transgenic mice has been found to be normal (Przedborski et al. 1992).

What we have done with CuZnSOD in the mouse can, of course, be done with any gene of interest, and a vast variety of different types of transgenic mice have already been made. However, it must always be kept in mind that, although it is likely that the results obtained in such mice will generally be applicable to humans, significant species differences between mouse and man may exist. Therefore, independent verification in humans of the results obtained with mouse models will ultimately have to be obtained.

Acknowledgments. This work was supported by grants from the National Institutes of Health (AG-08938, NS-14543, NS-25372), the American Cancer Society (CN-2), and the Parkinson's Disease Foundation.

References

Avraham KB, Schickler M, Sapoznikov D, Yarom R, Groner Y (1988) Down's syndrome: Abnormal neuromuscular junction in tongue of transgenic mice with elevated levels of human Cu/Zn-superoxide dismutase. Cell 54:823–829

Avraham KB, Sugarman H, Rotshenker S, Groner Y (1991) Down's syndrome: morphological remodelling and increased complexity in the neuromuscular junction of transgenic CuZn-superoxide dismutase mice. J Neurocytol 20:208–215

Cadet JL, Katz M, Jackson-Lewis V, Fahn S (1989) Vitamin E attenuates the toxic effects of intrastriatal injection of 6-hydroxydopamine (6-OHDA) in rats: behavioral and biochemical evidence. Brain Res 476:10–15

Ceballos-Picot I, Nicole A, Briand P, Grimber G, Delacourte A, Defossez A, Javoy-Agid F, Lafon M, Blouin JL, Sinet PM (1991) Neuronal-specific expression of human copper-zinc superoxide dismutase gene in transgenic mice: animal model of gene dosage effects in Down's syndrome. Brain Res 552:198–214

Chan PH (1989) The role of oxygen free radicals in brain injury and edema. In: Chow CK (ed) Cellular antioxidant defense mechanisms. CRC Press, Boca Raton FL, pp 89–109

Chan PH, Longar S, Fishman RA (1987) Protective effects of liposome-entrapped superoxide dismutase on post-traumatic brain edema. Ann Neurol 21:540–547

Chan PH, Chu L, Chen SF, Carlson EJ, Epstein CJ (1990a) Attenuation of glutamate-induced neuronal swelling and toxicity in transgenic mice overexpressing human CuZn-superoxide dismutase. Acta Neurochirurgica Suppl 51:245–247

Chan PH, Chu L, Chen SF, Carlson EJ, Epstein CJ (1990b) Reduced neurotoxicity in transgenic mice overexpressing human copper-zinc-superoxide dismutase. Stroke 21 (suppl III):III-80-III-82

Chan PH, Yang GY, Chen SF, Carlson E, Epstein CJ (1991) Cold-induced brain edema and infarction are reduced in transgenic mice ovreexpressing CuZn-superoxide dismutase. Ann Neurol 29:482–486

Choi DW (1990) Methods for antagonizing glutamate neurotoxicity. Cerebrovasc Brain Metab Rev 2:105–147
Cohen MV (1989) Free radicals in ischemic and reperfusion myocardial injury: is this the time for clinical trials? Ann Int Med 111:918–931
Costatini F, Chada K, Magram J (1986) Correction of murine β-thalassemia by gene transfer into the germ line. Science 233:1192–1194
Das DK, Russell JC, Jones RM (1991) Reduction of cold injury by superoxide dismutase and catalase. Free Rad Res Comm 12–13: 653–662
Dauber IM, Lesnefsky EJ, VanBenthuysen KM, Weil JV, Horwitz LD (1991) Reactive oxygen metabolite scavengers decrease functional coronary microvascular injury due to ischemia-reperfusion. Am J Physiol 260:H42–H49
Droy-Lefaix MT, Drouet Y, Geraud G, Hosford D, Braquet P (1991) Superoxide dismutase (SOD) and the PAF-antagonist (BN 52021) reduce small intestinal damage induced by ischemia-reperfusion. Free Rad Res Comm 12–13:725–735
Epstein CJ, Avraham KB, Lovett M, Smith S, Elroy-Stein O, Rotman G, Bry C, Groner Y (1987) Transgenic mice with increased CuZn-superoxide dismutase activity: Animal model of dosage effects in Down syndrome. Proc Natl Acad Sci USA 84:8044–8088
Groner Y, Avraham KB, Schickler M, Yarom R, Knobler H (1990) Clinical symptoms of Down syndrome are manifested in transgenic mice overexpressing the human CuZn-superoxide dismutase gene. In: Patterson D, Epstein CJ (eds) Molecular genetics of chromosome 21 and Down syndrome. Wiley-Liss, New York, pp 233–262
Hall ED, Braughler JM (1989) Central nervous system trauma and stroke. II. Physiological and pharmacological evidence for involvement of oxygen radicals and lipid peroxidation. Free Radic Biol Med 6:303–313
Halliwell B, Gutteridge JMC (1989) Free radicals in biology and medicine, 2nd ed, Oxford University Press, New York
Huang T-T, Carlson EJ, Leadon SA, Epstein CJ (1992) Relationship of resistance to oxygen free radicals to CuZn-superoxide dismutase activity in transgenic, transfected, and trisomic cells. FASEB J 6:903–910
Imaizumi S, Woolworth V, Fishman RA, Chan PH (1990) Liposome-entrapped superoxide dismutase reduces cerebral infarction in focal cerebral ischemia. Stroke 21: 1312–1317
Jackson RM, Veal CF, Beckman JS, Brannen AL (1990) Polyethylene glycol-conjugated superoxide dismutase in unilateral lung injury due to re-expansion (re-oxygenation). Am J Med Sci 300:22–28
Javitch JA, D'Amato RJ, Strittmatter SM, Snyder SH (1985) Parkinsonism-inducing neurotoxin, N-methyl-4-phenyl-1,2,3,6-tetrahydropyridine: Uptake of the metabolite N-methyl-4-phenylpyridine by dopamine neurons explains selective toxicity. Proc Natl Acad Sci USA 82:2173–2177
Jenner P (1989) Clues to the mechanisms underlying dopamine cell death in Parkinson's disease. J Neurol Neurosurg Psychiat Spec Suppl:22–28
Kinouchi H, Epstein CJ, Mizui T, Carlson E, Chen SF, Chan PH (1991) Attentuation of focal cerebral ischemic injury in transgenic mice overexpressing CuZn-superoxide dismutase. Proc Natl Acad Sci USA 88:11158–11162
Koke JR, Christodoulides NJ, Chudej LL, Bittar N (1990) Exogenous superoxide dismutase and catalase promote recovery of function in isolated rat heart after regional ischemia and may be transported from capillaries into myocytes. Mol Cell Biochem 96:97–105
Kontos HA (1985) Oxygen radicals in cerebral vascular injury. Circ Res 57:508–516
Krishnamoorthy G, Hinkle PC (1988) Studies on the electron transfer pathway topography of iron-sulfur centers, and site of coupling in NADH-Q oxidoreductase. J Biol Chem 263:17566–17575
Kujirai K, Jackson-Lewis V, Fahn S, Carlson E, Epstein CJ, Cadet JL (1991) Comparative quantitative autoradiographic distribution of μ opiod receptors in the brains of superoxide dismutase-transgenic (SOD-Tg) mice and of their non-transgenic (Non-Tg) littermates: age effects (abstract). Soc Neurosci Abstr, 17:812
Kyle JW, Birkenmeier EF, Gwynn B, Vogler C, Hopper PC, Hoffmann JW, Sly WS (1990) Correction of murine mucopolysaccharidosis VII by a human α-glucuronidase transgene. Proc Natl Acad Sci USA 87:3914–3918

Liu TH, Beckman JS, Freeman BA, Hogan EL, Hsu CY (1989) Polyethylene glycol-conjugated superoxide dismutase and catalase reduce ischemic brain injury. Am J Physiol 256:H589–H593

Markey SP, Johannessen JN, Chiueh CC, Burns RS, Herkenham MA (1984) Intraneuronal generation of a pyridinium metabolite may cause drug-related Parkinsonism. Nature 311:464–467

Przedborski S, Kostic V, Jackson-Lewis V, Carlson E, Epstein CJ, Cadet JL (1991) Quantitative autoradiographic distribution of [^{3}H]-MPTP binding in the brains of superoxide dismutase transgenic mice. Brain Res Bull 26:987–991

Przedborski S, Kostic V, Jackson-Lewis V, Naini AB, Simonetti S, Fahn S, Carlson E, Epstein CJ, Cadet JL (1992) Transgenic mice with increased Cu/Zn-superoxide dismutase activity are resistant to 1-methyl-4-phenyl-1,2,3,6-tetrahydropyridine-induced neurotoxicity. J Neurosci 12:1658–1667

Rossetti ZL, Sotgiu A, Sharp DE, Hadjiconstantinou M, Neff NH (1988) 1-Methyl-4-phenyl-1,2,3,6-tetrahydropyridine (MPTP) and free radicals in vitro. Biochem Pharmacol 37:4573–4574

Sinet P-M (1982) Metabolism of oxygen derivatives in Down's syndrome. Ann NY Acad Sci 396:83–94

Sinet P-M, Lejeune J, Jerome H (1979) Trisomy 21 (Down's syndrome), Glutathione peroxidase, hexose monophosphate shunt and IQ. Life Sci 24:29–33

Spina MB and Cohen G (1989) Dopamine turnover and glutathione oxidation: implications for Parkinson disease. Proc Natl Acad Sci USA 86:1398–1400

Wang J, Chen H, Wang T, Diao Y, Tian K (1990) Oxygen-derived free radicals induced cellular injury in superior mesenteric artery occlusion shock: protective effect of superoxide dismutase. Circ Shock 32:31–41

Gene Transfer into the Nervous System using Recombinant Herpes Virus Vectors

X. O. Breakefield, Q. Huang, J. K. Andersen, M. F. Kramer, W. R. Bebrin, G. Davar, B. Vos, D. A. Garber, M. DiFiglia and *D. M. Coen*

Abstract

Gene delivery to the adult rodent brain can be achieved using recombinant virus vectors derived from herpes simplex virus type 1 (HSV-1). This double stranded DNA virus has a large genome (150 kb) which may carry up to 50 kb of foreign DNA at replaceable sites. This virus has a number of advantages for delivery of genes to neurons. It can be taken up by nerve terminals and transported by rapid retrograde transport to the cell nucleus; it can be passed transsynaptically between neurons; and it can enter latency in many types of neurons and possibly other cell types. In latency, the virus exists in a stable and essentially benign form in the nucleus, with active transcription predominantly from one viral promoter. Our group and others have generated vectors which are replication-compromised or replication-defective and have diminished pathogenicity to animals. These vectors can confer both short- and long-term expression of foreign genes on neurons in the peripheral and central nervous systems. Examples will be given here for gene delivery to the adult rat nervous system using vectors which lack viral thymidine kinase activity and cannot replicate in neurons, focusing on the use of two non-herpes virus promoters – the Moloney murine leukemia virus (MoMLV) long terminal repeat (LTR) promoter and the rat neuronal specific enolase promoter – to confer expression of the *E. coli lacZ* gene on neural cells in the brain. Herpes vectors can potentially be used to change the physiology of specific sets of endogenous neurons following stereotactic inoculation into the rat brain. This gene delivery scheme might provè useful in therapeutic paradigms where subsets of neurons need to be reprogrammed to augment their survival or function.

Introduction

In designing therapies for neurologic diseases it would be very valuable to have a means of introducing genes into the adult nervous system. This would allow manipulation of the physiology of the nervous system in a controllable manner. There are three current methods for introducing genes into focal regions of the nervous system through stereotactic injections: retrovirus vectors, grafted cells, and herpes virus vectors. Retrovirus vectors have several disadvantages for the adult nervous system. Since they preferentially introduce genes into the genome of cells undergoing DNA replication, the virus particles have a short half-life, and relatively little foreign DNA (~ 8 kb) can be carried by the vector (Miller et al. 1990; Cepko 1988). A variety of cell types, with and without prior genetic modification, can be grafted into the brain as a means to release or metabolize neuroactive compounds (Gage and Fisher 1991). Although these grafts can be retained for extended periods, most will eventually be rejected and are thus not "permanent." So far the only means to effect stable gene transfer to neurons is through the use of herpes virus vectors (for additional review see Roizman and Jenkins 1985; Barker and Roizman 1990; Breakefield and De Luca 1991). In this chapter we will review the progress to date in the development of recombinant virus vectors for gene delivery *in vivo*.

Herpes virus vectors for gene transfer to the nervous system

Herpes virus, in particular herpes simplex virus type 1 (HSV-1), have several features that make them well suited for gene delivery to the nervous system (Table 1; for review see Breakefield and DeLuca 1991). This DNA virus is taken up by essentially all cells with varying degrees of efficiency. In neurons it can be taken up anywhere along the cell surface (Lycke et al. 1988). The enveloped virus fuses with the cell membrane releasing the virus capsid into the cytoplasm. In neurons, the virus capsid associates with the microtubular assembly responsible for rapid retrograde transport and is efficiently carried by fast axonal transport from the nerve terminals to the cell bodies (Lycke et al. 1984). The capsid, in turn, is transported to the nucleus where the viral DNA is released intact. In most cells the virus replicates within the nucleus, eventually killing the cell. In neurons, the virus may replicate and be passed by anterograde transport to the presynaptic terminals, where it can cross the synapse to postsynaptic neurons and thus progress through the nerve network, killing neurons along its route (Ugolini et al. 1989; Card et al. 1990; La Vail et al. 1990). In some cases it may enter a latent state where very few of the viral genes are expressed and the viral DNA evidently exists as an episomal element in the nucleus (Stevens 1989). In this state the virus is apparently benign, but may be reactivated to a replicative state by a variety of stimuli which alter the physiology of the neuron. It is not clear whether reactivation causes death of the host neuron.

Several questions must be addressed in order to develop HSV-1 recombinant virus vectors for gene delivery to neurons. First, where can genes be introduced

Table 1. HSV-1 as vectors

- Genome can carry multiple inserts, up to 50 kb.
- Mutant viruses have reduced pathogenicity.
- Virus travels by retrograde transport within neurons and crosses synapses.
- Virus can enter latency in some neurons (benign).
- Some promoters in virus are active in latency and can confer stable gene expression.

into the herpes genome and which promoters should be used to achieve gene expression in latency? Second, how can viral toxicity be reduced? Third, which cells will harbor latency and how can the virus be "steered" into this state? Each of these issues will be addressed in terms of vectors generated by ourselves and others.

Viral vectors gene delivery

The HSV-1 genome is about 150 kb in length and contains 72 genes distributed in unique and repeat regions (Fig. 1; for review see Roizman and Sears 1990). Extensive mutational analysis has revealed that large segments of the genome (probably up to 50 kb) can be replaced at multiple sites without markedly compromising the ability of the virus to replicate in at least some cells (Roizman and Jenkins 1985; Barker and Roizman 1990; Schranz et al. 1989). A number of recombinant vectors have been described which can confer transient expression of marker genes on neurons of the central and peripheral nervous system in rodents (for example see Table 2).

A viral promoter element has been described that regulates expression of latency associated transcripts (LATs; Stevens et al. 1987). Foreign genes, *E. coli LacZ* and human beta-globin, placed downstream of this promoter are expressed stably in sensory neurons of the mouse following inoculation of the virus at peripheral sites (Dobson et al. 1989; Ho and Mocarski 1989). Since LAT sequences lie in a repeat region and, as a consequence, are diploid, the foreign gene must be introduced into both sites to ensure stability of the vector, which complicates construction of the recombinant vectors to some extent. Disruption of the viral sequence encoding the LATs does not appear to interfere with the ability of the virus to enter latency, but may compromise reactivation *in vivo* (Hill et al. 1990; Leib et al. 1989; Steiner et al. 1989). Stable expression of beta-galactosidase in sensory neurons was also observed in virus where *lacZ* was downstream of the MoMLV LTR, a strong, constitutive promoter in most cells (Dobson et al. 1990).

Table 2. Examples of other recombinant HSV-1 vectors for gene delivery to neurons *in vivo*.

Foreign Gene	Type of Virus Mutant	Length of Neuron	Expression	Reference
pICPO-lacZ	ICPO$^-$	CNS	transient	Cai and Schaffer 1989: Chiocca et al. 1990; Huang et al. 1992
pLAT-beta globin	LAT$^-$	sensory	long term	Dobson et al. 1989
"pLAT"-lacZ	LAT$^-$	sensory	long term	Ho and Mocarski 1989
pMoMLVLTR-lacZ	LAT$^-$, ICP4$^-$	sensory & motor	long term	Dobson et al. 1990
pgC-lacZ	US3$^-$	CNS	transient	Fink et al. in press

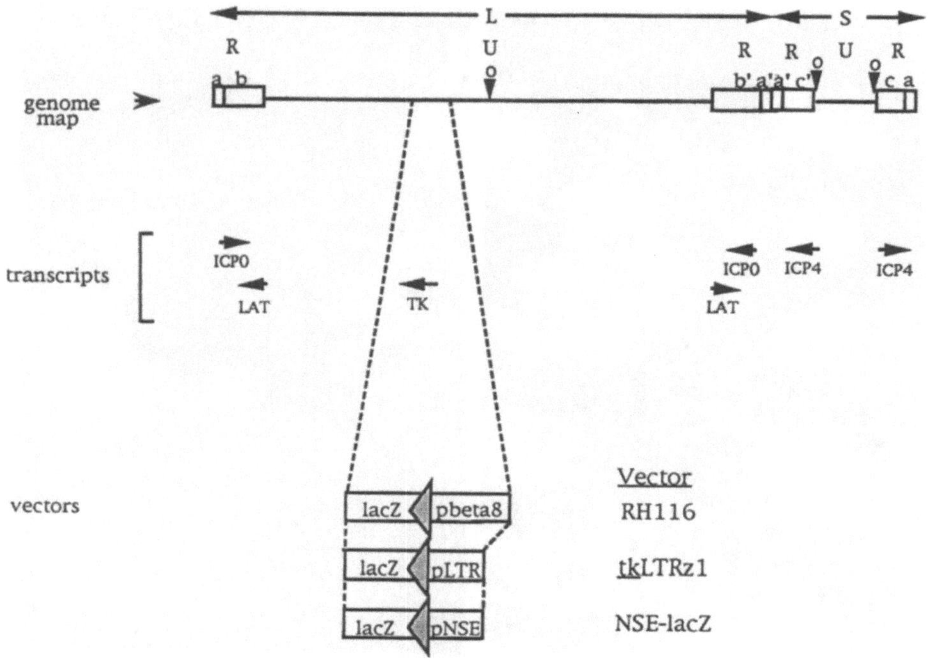

Fig. 1. Diagram of HSV-1 genome and recombinant vectors. Map units (m.u.) are shown in the top line along the approximately 150 kb genome. Repeat elements (R_L and R_S) contain repeats a, b and a, c, respectively. Unique regions (U) are designated as long (L) and short (S). Three origins or replications (O) are designated by arrow heads. The position of a few of the viral transcripts discussed in this paper are indicated by horizontal arrows. (Modified from and reproduced with permission of McGeoch et al. 1990). The three TK-minus viral vectors described in this paper are indicated at the bottom giving the promoter driving the lacZ gene. RH 116, beta 8 promoter (Ho and Mocarski 1988), *tk*LTRz1, MoMLVLTR promoter (Davar et al. 1991); and NSE-lacZ, NSE promoter (Andersen et al., in press).

In our studies, relatively long-term (2–3 weeks) expression of the *lacZ* gene has been observed in adult rat brain with vectors where it is placed under the control of either the MoMLV LTR or the neuronal specific enolase (NSE) promoter within the HSV-1 thymidine kinase (TK) locus (Fig. 1). There are several advantages to introducing foreign genes at the viral TK site: there is only one site present in the genome; disruption of the TK gene allows the recombinant virus to be selected by growth in the presence of nucleoside analogues, such as acyclovir; loss of viral TK activity renders the virus replication-defective in non-dividing cells; and TK-minus viruses can enter latency, but can not reactive from it (Coen et al. 1989; Efstathiou et al. 1989). Expression of *lacZ* under control of the LTR promoter was monitored of 3 and 17 days following stereotactic inoculation of 1.8 × 10^5 pfu of the *tk*LTRz1 vector in 1 μl into the caudate-putamen of adult rats (Fig. 2). Similar results were noted with histochemical and immunocytochemical staining unless otherwise indi-

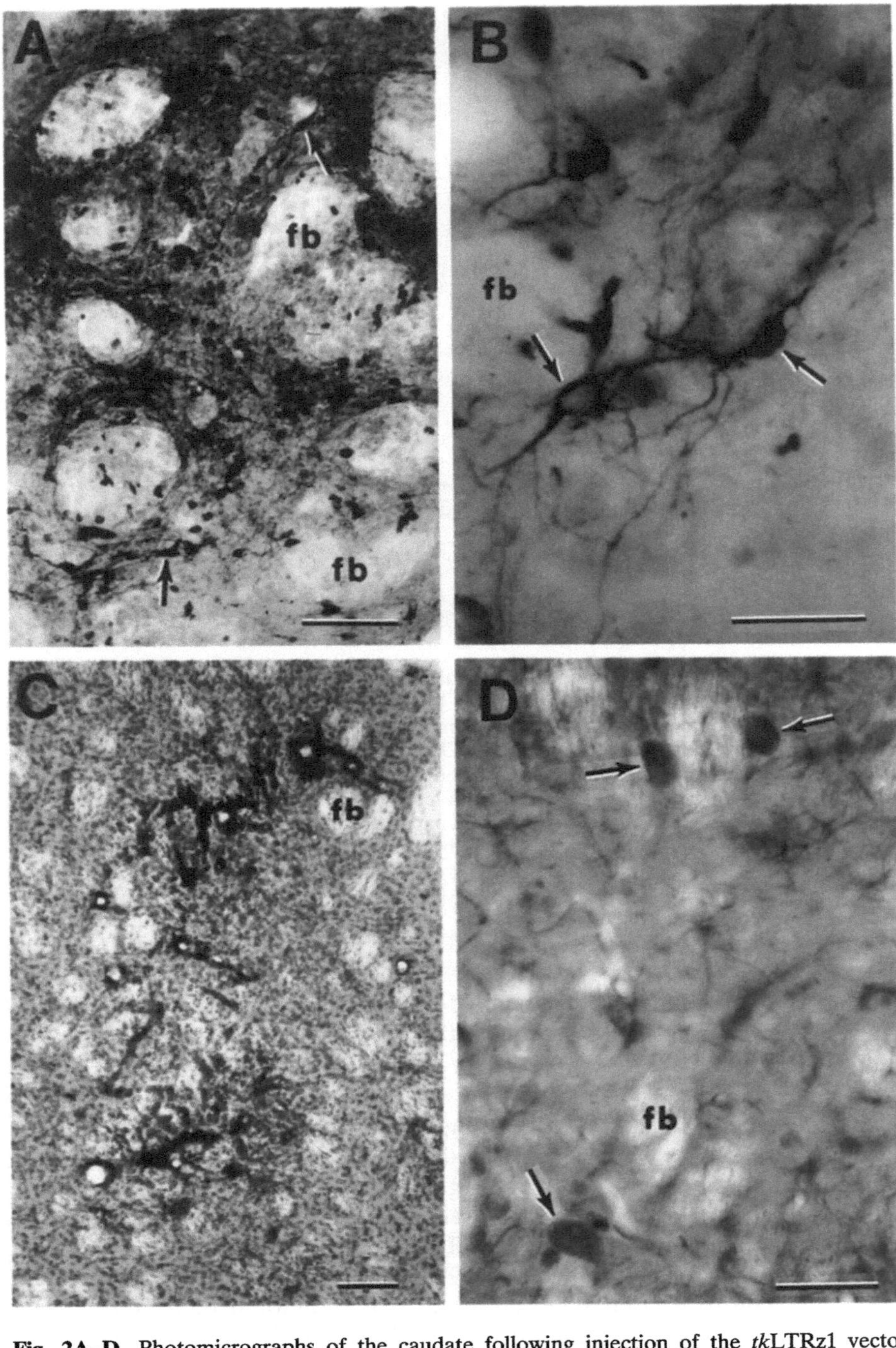

Fig. 2A–D. Photomicrographs of the caudate following injection of the *tk*LTRz1 vector. Adult rats were injected stereotactically with 2×10^5 pfu of the virus vector in 1 μl into the neostriatum. Following perfusion and sectioning, sections of brain tissue around the injection

cated. At three days, immunohistochemical staining using polyclonal antibodies specific to *E. coli* beta-galactosidase revealed extensive labeling of medium-sized and large striatal neurons (identified by their characteristic morphology), and some labeling of astrocytes, with very little tissue damage around the injection site (Fig. 2A and B). After 17 days, histochemical staining for *lacZ* expression, using the X-gal substrate (Turner and Cepko 1987) combined with a hematoxylin and eosin counterstain, revealed labeling of mostly small (< 10 μm in diameter), apparently non-neuronal cells (Fig. 2C). At 17 days some immunoreactive glia and a few large neurons (> 20 μm in diameter) were also apparent with immunohistochemical staining, which is more sensitive than histochemical staining (Huang et al. 1992; Fig. 2D). Although there was no evidence of atrophy in the caudate-putamen itself after 17 days, in one animal there was extensive gliosis and disruption of the corpus callosum overlying the caudate-putamen, suggesting some viral replication and/or an immune response to the virus (Fig. 2C). The HSV-1 TK-minus vectors are relatively non-pathogenic, however (see also Huang et al. 1992). The LTR promoter in this context behaves differently than does an immediate-early HSV promoter inserted at the same site in the viral genome. When the TK-minus vector, RH-116 (Ho and Mocarski 1988), which bears the *lacZ* gene under the viral beta-8 promoter, was inoculated into adult rat brain, expression was noted only after three days, and not after 14 days (Chiocca et al. 1990; Andersen, unpublished data). Although expression from the LTR extends to at least 17 days in rat brain, it does not appear to be "stable", as no positively staining cells were noted at the injection site 28 days after inoculation, the next time point examined. Either this vector is not entering latency in cells in this brain region, or the LTR promoter is not stably active in latency at this site in the herpes genome. Expression of the *lacZ* gene driven by the NSE promoter appeared to be lower than that with the LTR in these TK-minus vectors, as fewer, less intensely labeled cells were observed three days after injection into the caudate-putamen, and no definitively *lacZ* positive cells were observed after 17 days (data not shown). In contrast, when adult rats were injected with the NSE-lacZ vector in the cerebral cortex, intense histochemical labeling of a few neuronal appearing cells was observed 14 and 30 days after inoculation at sites some distance from the injection, suggesting retrograde transport (Fig. 3A; Andersen et al., in press). This selective expression pattern may reflect the fact that this enzyme is present at high levels in pyramidal neurons in the cortex

---

site were stained for beta-galactosidase. *A,* immunohistochemical staining using polyclonal antibodies for *E. coli* beta-gal (5 Prime - 3 Prime) with nickel intensification. Extensive labeling of immunoreactive beta-galactosidase is present in medium-sized and large neurons within and around the needle tract. Some small, intensely labeled cells are probably glia. The tissue damage caused by this mutant virus is minimal. Arrows indicate striatal neurons; fb, fiber bundle. Scale bar, 100 μm. *B,* a higher magnification of the same area shown in (A). The labeled fine processes are clearly visible. Scale bar, 50 μm. *C,* animals were sacrificed 17 days after injection and stained for beta-galactosidase histochemically (Turner and Cepko 1987) with hematoxylin and eosin counterstaining. "Blue" labeling appears only in small non-neuronal cells within the injection site. Scale bar, 100 μm. *D,* immunocytochemical staining for beta-glactosidase 17 days after inoculation. Arrows indicate cell bodies (> 20 μm in diameter) of large striatal neurons about 200 μm lateral to the injection site. Scale bar, 50 μm.

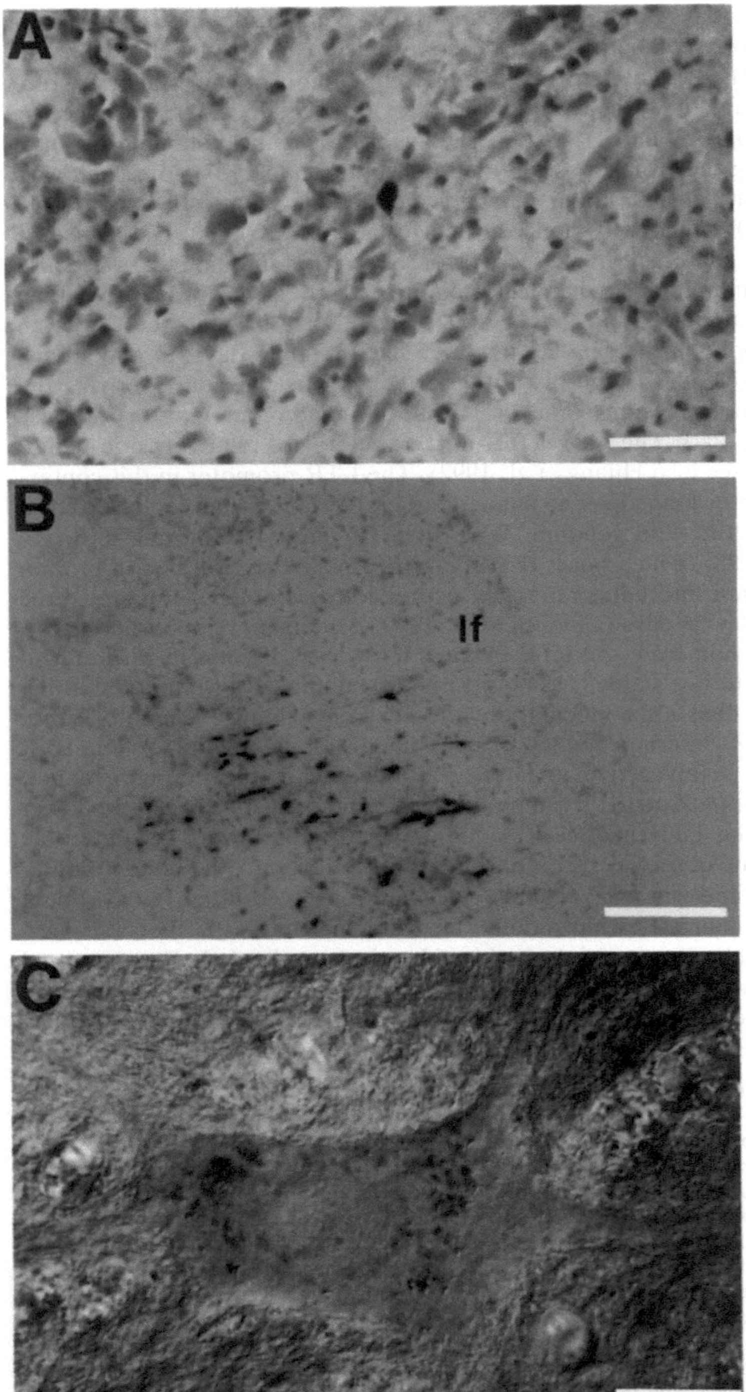

Fig. 3 A–C.

suggesting strong NSE promoter activity in these cells (Marangos and Schmechel 1987).

Although long-term expression of the marker gene may represent transcription from a latent viral genome, it is also possible that the TK-minus vectors continue to replicate at a low level in some dividing cells in the brain, such as glia, ependymal cells or endothelial cells, and that occasional expression after two weeks reflects a recent infection of these cells. This finding of continued expression must be confirmed by additional means, as occasional neuronal-like cells positive for *lacZ* by histochemical analysis have been observed in control rat brains even under conditions believed to be specific for the bacterial beta-galactosidase (Rosenberg et al. 1991; Emson et al. 1990), although immunocytochemical staining appears more specific (Rosenberg et al. 1991). Also it is important to realize that, in many regions of the brain, neurons cannot be defined by morphology alone. For example, short-term expression of *lacZ* in neuronal-like cells was observed in the lateral funiculus of the adult rat spinal cord following injection of a TK-minus, *lacZ* vector, RH116 (Ho and Mocarski 1988; 1.0×10^3 pfu in 5 μl) into the C7 segment (Fig. 3B). These cells are probably not neurons based on their position outside the spinal cord gray matter and the absence of visible nucleoli (seen at higher power, not shown), which are usually visible in neurons following hematoxylin-eosin counterstaining.

From all studies to date it appears possible that HSV-1-derived recombinant virus vectors can achieve long-term (2–3 weeks) or stable (> 1 month) expression of foreign genes in peripheral and central neurons when foreign genes are inserted in at least three different sites in the herpes genome (LAT, ICP4 and TK) using several different promoter elements, including the herpes LAT promoter, a promoter from another virus, MoMLV LTR, and a mammalian cell-specific promoter, pNSE. However, caution must be used in defining stability, authenticity and cell specificity of gene expression.

Fig. 3A–C. Virus-mediated delivery of lacZ to different cellular populations in the rat. Vectors were inoculated into three different areas of adult rats. After sacrifice and sectioning *lacZ* expression was evaluated by histochemical staining with X-gal, *A*, 10^6 pfu of the NSE-lacZ vector was injected in 10 μl into the frontal cortex. Animals were evaluated 14 days after inoculation. At some distance from the injection site, densely blue cells were noted which had the morphological features of pyramidal neurons. There was no apparent necrosis in this area. Scale bar, 50 μm. *B*, RH116 (1×10^3 pfu in 5 μl) was injected into one side of the rat dorsal spinal cord at the C7 level. Sections were counterstained with hematoxylin and eosin. lf; lateral funiculus. Scale bar, 120 μm. *C*, Photomicrograph of a neuron in the medullary dorsal horn showing a punctate staining pattern seven days following inoculation of the 7134 vector (2×10^5 pfu in 10 μl) on the scarified cornea of one rat eye. Seven days later animals were sacrificed and brainstem sections (40 μm) were taken through the trigeminal nucleus caudalis and stained for cytochemical detection of beta-galactosidase. The neuron shown in the photograph was located in the most ventral part of lamina IV of the medullary dorsal horn ipsilateral to the inoculated cornea. Note the punctate cytoplasmatic staining surrounding the cell nucleus. Scale, bar 10 μm.

Toxicity of herpes virus

HSV-1 virus is toxic to cells through a variety of mechanisms. Productive HSV infection resulting in viral replication generally kills cells. The ability of the virus to replicate in neurons is presumably controlled by a number of factors. These include the number of virions that enter the cell (multiplicity of infection, MOI), the type of neuron it enters and its transcriptional state, the species infected, and the strain of virus. In general, any of the following conditions will decrease the tendency of the virus to replicate in neurons: low MOI, non-traumatized cells, cell types that are permissive for latency, and attenuated strains of HSV. In addition, the ability of the virus to replicate can be controlled through mutations in the viral genome. Some mutations, e.g., ICP4-minus, make the virus unable to replicate in all cells except those genetically engineered to replace the missing function (DeLuca et al. 1985). Other mutations, e.g., TK-minus, will render the virus capable of replicating only in some cell types, i.e., in dividing cells where the deoxynucleotide triphosphate levels are high and evidently can replace the missing viral function. Still other mutations can render the virus replication-compromised in some or all cells, such as ICP0-minus (Cai and Schaffer 1989) and US3-minus (Fink et al. 1992), or can reduce its neurovirulence, e.g., gamma 34.5-minus (Chou et al. 1990). Drugs such as acyclovir can also be used to arrest viral replication of TK-positive virus, and other drugs such as foscarnet can be used for TK-minus virus (Martuza et al. 1991; Coen 1992).

Herpes simplex virus can also be toxic to cells even under conditions where it does not replicate. This is due to a complex set of functions which is not completely understood and appears to involve viral coat proteins as well as other viral proteins expressed during replicative infection. Some reduction in toxicity has been achieved through mutations in the VP16 coat protein, a very powerful transactivating factor (Ace et al. 1989), and in UL41, which encodes a protein that leads to degradation of host cellular RNA (Kwong and Frenkel 1989). So far no vectors have been reported which include deficiencies in these toxic functions.

Long-term expression of foreign genes has been described in sensory, as well as motor, neurons in the mouse with no apparent toxicity to these cells, even using vectors with full replicative potential (Dobsen et al. 1989 and 1990; Ho and Mocarski 1989). This may be due to the fact that few virus particles are taken up by retrograde transport from the nerve terminals and that these cells can harbor the virus in latency. In contrast, following inoculation of TK-minus vectors into rat brain, there was notable necrosis and cell degeneration at the site of injection, even when there was no apparent changes in the behavior or health of the animals (Huang et al. 1992). Injection of a replication-compromised, ICP0-minus vector (7134; Cai and Schaffer 1989) into rat brain caused a "smoldering" encephalitis such that the animals appeared healthy after two months, but there was extensive necrosis of brain tissue spreading out from the injection site (Huang et al. 1992). The advantage of the replication-compromised, as compared to replication-defective, vectors in the brain was that the former delivered the gene to more cells (Chiocca et al. 1990). Transsynaptic transfer of the *lacZ* gene from primary afferent terminals to central neurons was achieved within the trigeminal system using a replication-compromised vector. Seven days following inoculation of the ICP0-minus vector (7134; Cai and Schaffer 1989) onto the scarified cornea of the rat, several neurons in the ipsilateral medullary dorsal horn showed a punctate staining pattern for beta-galactosidase (Fig. 3C). These neurons were distributed at the midpoint in the rostrocaudal extent of the trigeminal nucleus caudalis, primarily in ventral lamina IV, close to the border with lamina V. This is the area where primary afferent neurons from the opthalmic division of the trigeminal nerve terminate. No positively staining neurons were found in the medullary dorsal horn, contralateral to the inoculated cornea. The size and location of the labeled cells suggest that they are projection neurons of ascending somatosensory pathways.

It is clear that decreasing the toxicity of herpes remains a critical issue in the development of viral vectors for neurons. Given the current state of the art one can expect focal damage in the brain at the site of the injection, although this can be very limited. Since neurons interact in complex networks, it can also be speculated that HSV-mediated death of some neurons could lead to eventual death of other neurons which interact with them, causing the collapse of the network, as proposed for neurodegenerative diseases (Roberts and DiFiglia 1989). To maintain the integrity of neurons targeted for gene delivery, it is probably best to inject the virus near the nerve terminals and let it be delivered by retrograde transport back to the cell nucleus. If some degeneration of the nervous system can be tolerated in an experimental system, it may be possible to deliver a foreign gene to large numbers of cells throughout the brain using replication-compromised vectors, whose replication can be stopped by drug treatment.

Host cells that can harbor latency

The state of latency has been described most thoroughly for the sensory neurons of the mouse and rabbit (Stevens 1989). The factors that promote latency and the viral and host genes that mediate it are not understood, however. In general, it is believed that replication is not necessary for latency to occur, and that decreased expression of immediate-early viral genes promotes latency (Kemp et al. 1990; Stevens 1989; Katz et al. 1990; Steiner et al. 1990; Sears et al. 1991; Kosz-Vnenchak et al. 1990). The question of which cell types in the animal can harbor latency has also not been resolved completely. Latent HSV-1 has been demonstrated in a number of tissues, including brain and adrenal medulla (Stevens 1978). Within the brain it is not clear if all neurons or only some types of neurons and/or glia can harbor latent virus, although a number of cell types appear to be labeled by *in situ* hybridization to LATs (Deatly et al. 1988).

Stable gene expression using recombinant herpes virus vectors may be limited to cells which normally harbor latency, unless factors are discovered which can induce this state in other non-dividing cells. It may be possible to design vectors with an increased tendency to enter latency by blocking the induction of immediate-early viral genes, for example through mutations in transactiving viral proteins, such as VP16 (Steiner et al. 1990; Sears et al. 1991) and perhaps ICP4 (Katz et al. 1990), or in the TK gene needed for viral replication in neurons (Kosz-Vnenchak et al. 1990). Extensive studies will be needed to determine which cells can harbor latency. The presence of latent virus is indicated in a number of ways: reactivation of the virus following explantation of the tissue and culturing with permissive cells, PCR amplification of viral DNA (Katz et al. 1990), *in situ* hybridization to LATs in the cell nucleus (Rock et al. 1987; Stevens et al. 1989), or demonstration of foreign marker gene expression driven by the LAT promoter (Ho and Mocarski 1989). All these analyses, except *in situ* hybridization to a probe that recognizes a unique portion of the LATs, are complicated by the fact that, unless the vector is completely replication-defective in all cells, long-term expression could be due to "smoldering" infections in one cell type with continual re-infection of neurons and temporary expression of foreign genes in them. Also LATs are present at low levels during the replicative cycle (Dobson et al. 1989). The best evidence to date for stable gene delivery to neurons is the work by Dobson et al. 1990) in which a LAT-defective, ICP4-minus replication defective vector conferred expression of *lacZ* apparently under the control of the MoMLVLTR for 24 weeks in sensory neurons *in vivo*.

Model systems for gene delivery using herpes virus vectors

Many ongoing studies are directed toward working out the "bugs" of the herpes vectors to reduced viral toxicity and achieve stable foreign gene expression in some cells in the nervous system of experimental animals. The next goal will be to use these vectors to understand the function of the nervous system and to effect therapeutic intervention. In considering this goal it is useful to consider possible modes of delivery. Vectors can be inoculated at peripheral nerve terminals, for example in the skin or cornea, or they can be stereotactically introduced into specific regions of the brain (Fig. 4). If vectors are replication-defective, gene delivery will be limited to those cells which have at least part of their surface in contact with the initial inoculation. These vectors can probably be made very "safe" by deleting several viral genes necessary for replication (e.g., ICP4, ICP27), so that there is essentially no chance of a replication-competent, revertant virus being generated *in vivo*. If the vectors are replication-compromised (e.g., ICP0-minus, US3-minus), the virus has the potential to cross synapses to other neurons and to spread, presumably slowly and in preferred cell types, throughout the brain. Such vectors may kill some cells, but at the same time they may effect gene delivery to a large number of cells. Cell specificity of expression may be achieved through the use of cell-specific promoters to drive expression of foreign genes. It may be possible to "impede" replication of these vectors through drug treatment or conditional mutations. If the vectors could be "encouraged" to replicate in astrocytes and enter latency in neurons, a controllable spread of the virus might be achieved

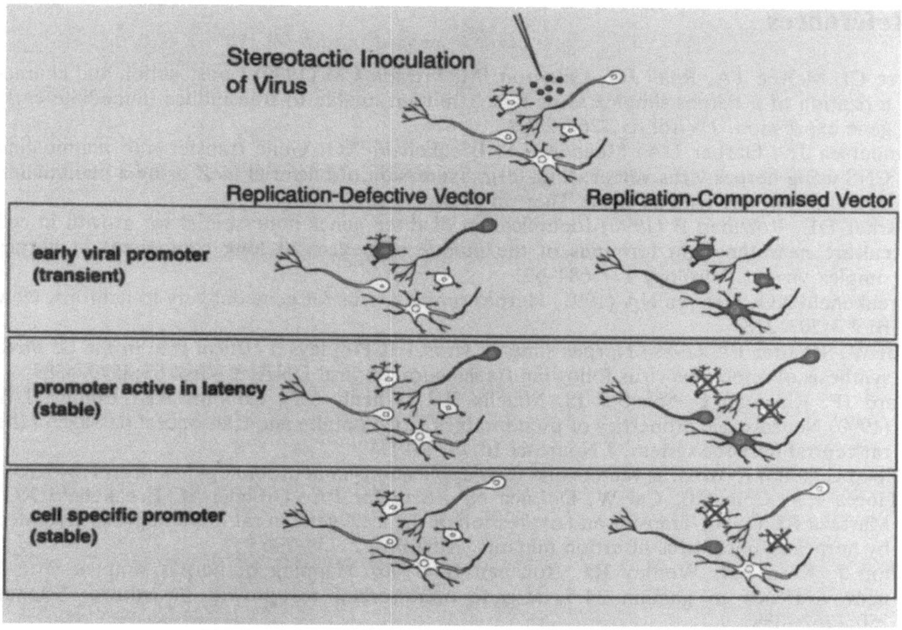

Fig. 4. Diagrammatic representation of gene delivery to the brain. Replication-defective vectors deliver the gene only to cells in contact with the initial inoculum. Replication-compromised vectors can be passed to additional cells following replication in some cells, but viral replication will kill the host cell. Early viral promoters can effect transient gene expression in all infected cells. If the virus enters latency, stable expression can be achieved in all or a portion of the cells harboring latency, depending on the promoter used.

with only minimal damage to the brain, as the astrocyte population could be replenished through cell division. Clearly any viral replication would trigger an immune reaction, which could also have detrimental effects on the brain. In the case of the vectors described here, the TK-minus ones appear to exhibit very limited replication in the rat brain, with only focal necrosis at the site of injection and no apparent pathologic effects to the animals even after several months; the ICP0-minus vector causes progressive neuronal damage over time, although the animals appear healthy for months after intracerebral inoculation (Huang et al. 1992).

Acknowledgments. We thank Ms. Suzanne McDavitt for skilled preparation of this manuscript. This work was supported by NIH grant NS24279 (XOB); CIDA NS01497 (GD); NRSA F32 NS08810 (JA); NIH grants PO1 AI24010, S07 RR05381 (DC); NIH grant NS16367 (MD).

References

Ace CI, McKee TA, Ryan JM, Cameron JM, Preston CM (1989) Construction and characterization of a herpes simplex virus type 1 mutant unable to transinduce immediate-early gene expression. J Virol 63:2260–2269

Andersen JK, Garber DA, Meaney CA, Breakefield XO. Gene transfer into mammalian CNS using herpes virus vector: long-term expression of bacterial *lacZ* using a mammalian neural promoter. Human Gene Ther, in press

Barker DE, Roizman B (1990) Identification of three genes nonessential for growth in cell culture near the right terminus of the unique sequences of long component of herpes simplex virus 1. Virology 177:684–691

Breakefield XO, DeLuca NA (1991) Herpes simplex virus for gene delivery to neurons. New Biol 3:203–218

Cai W, Schaffer PA (1989) Herpes simplex virus 1 ICP0 plays a critical role in the de novo synthesis of infectious virus following transfection of viral DNA. J Virol 63:4579–4589

Card JP, Rinaman L, Schaber JS, Miaella RR, Whealy MF, Robbins AK, Enquist LW (1990) Neurotropic properties of pseudorabies virus: uptake and transneural passage in the rat central nervous system. J Neurosci 10:1974–1994

Cepko C (1988) Retrovirus vectors and their applications in neurobiology. Neuron 1:345–353

Chiocca EA, Chio BB, Cai W, DeLuca NA, Schaffer PA, DiFiglia M, Breakefield XO, Martuza RL (1990) Transfer and expression of the lacZ gene in rat brain neurons mediated by herpes simplex virus insertion mutants. New Biol 2:739–746

Chou J, Kern ER, Whitley RJ, Roizman B (1990) Mapping of herpes simplex virus-1 neurovirulence to gamma 34.5, a gene nonessential for growth in culture. Science 250:1262–1266

Coen DM, Kosz-Vnenchak M, Jacobsen JG, Leib DA, Bogart CL, Schaffer PA, Tyler KL, Knipe MD (1989) Thymidine kinase-negative herpes simplex virus mutants establish latency in trigeminal ganglia but do not reactivate. Proc Natl Acad Sci USA 86:4736–4740

Coen DM (1992) Molecular aspects of anti-herpes virus drugs. Sem Virol, in press

Davar G, Meaney C, Andersen JK, Kramer M, Bebrin W, Voss BP, Burstein R, Garber D, Rosenberg W, Maciewicz R, Coen D, Breakefield XO (1991) Gene delivery to rat sensory ganglia and spinal cord cells using herpes virus vectors. Soc Neuroscience Abst 399.6

Deatly AM, Spivack JG, Lavi E, O'Boyle DR II, Frase NW (1988) Latent herpes simplex virus type 1 transcripts in peripheral and central nervous system tissues of mice map to similar regions of the viral genome. J Virol 62:749–756

DeLuca NA, McCarthy AM, Schaffer PA (1985) Isolation and characterization of deletion mutants of herpes simplex virus type 1 in the gene encoding immediate-early regulatory protein ICP4. J Virol 56:558–570

Dobson AT, Sederati F, Devi-Rao G, Flanagan WM, Farrell MJ, Stevens JG, Wagner EK, Feldman LT (1989) Identification of the latency-association transcript promoter by expression of rabbit beta-globin mRNA in mouse sensory nerve ganglia latently infected with a recombinant herpes simplex virus. J Virol 63:3844–3851

Dobson AT, Margolis TP, Sedarati F, Stevens JG, Feldman LT (1990) A latent, nonpathogenic HSV-1 derived vector stably expresses beta-galactosidase in mouse neurons. Neuron 5:353–360

Efstathiou S, Kemp S, Darby G, Minson AC (1989) The role of herpes simplex virus type 1 thymidine kinase in pathogenesis. J Gen Virol 70:869–879

Emson PC, Shoham S, Feler C, Buss T, Price J, Wilson CJ (1990) The use of a retroviral vector to identify foetal striatal neurones transplanted into the adult striatum. Exp Brain Res 79:427–430

Fink DJ, Sternberg LR, Weber PC, Mata M, Goins WF, Glorioso JC (1992) In vivo expression of beta-galactosidase in hippocampal neurons by HSV-mediated gene transfer. Human Gene Ther 3:11–20

Gage FH, Fisher LJ (1991) Intracerebral grafting: A tool for the neurobiologist. Neuron 6:1–12

Hill JM, Sederati F, Javier RT, Wagner EK, Stevens JG (1990) Herpes simplex virus latent phase transcription facilitates in vivo reactivation. Virology 174:117–125

Ho DY, Mocarski ES (1988) Beta-galactosidase as a marker in the peripheral and neural tissues of the herpes simplex virus-infected mouse. Virology 167:279–283

Ho DY, Mocarski ES (1989) Herpes simplex virus latent RNA (LAT) is not required for latent infection in the mouse. Proc Natl Acad Sci USA 86:7596–7600

Huang Q, Vonsattel JP, Schaffer PA, Martuza RL, Breakefield XO, DiFiglia M (1992) Introduction of a foreign gene *(E. coli lacZ)* into rat neostriatal neurons using herpes simplex virus mutants: A light and electron microscopic study. Exp Neurol 115:303–316

Katz JP, Bodin ET; Coen DM (1990) Quantitative polymerase chain reaction analysis of herpes simplex virus DNA in ganglia of mice infected with replication-incompetent mutants. J Virol 64:4288–4295

Kemp LM, Dent CL, Latchman DS (1990) Octamer motif mediates transcriptional repression of HSV immediate-early genes and octamer-containing cellular promoters in neuronal cells. Neuron 4:215–222

Kosz-Vnenchak M, Coen DM, Knipe DM (1990) Restricted expression of herpes simplex virus lytic genes during establishment of latent infection by thymidine kinase-negative mutant viruses. J Virol 64:5396–5402

Kwong AD, Frenkel N (1989) The herpes simplex virus virion host shutoff function. J. Virol 63:4834–4839

LaVail JH, Zhan J, Margolis TP (1990) HSV (Type 1) infection of the trigeminal complex. Brain Res 514:181–188

Leib DA, Bogard CL, Kosz-Vnenchak M, Hicks KA, Coen DM, Knipe DM, Schaffer PA (1989) A deletion mutant of the latency-associated transcript of herpes simpex virus type 1 reactivates from the latent state with reduced frequency. J Virol 63:2893–2900

Lycke E, Kristensson K, Svennerholm B, Vahlne A, Ziegler R (1984) Uptake and transport of herpes simplex virus in neurites of rat dorsal root ganglia cells in culture. J Gen Virol 65:55–64

Lycke E, Hamark B, Johansson M, Krotochwil A, Lycke J, Svennerholm B (1988) Herpes simplex virus infection of the human sensory neuron. An electron microscopy study. Arch Virol 101:87–104

Marangos PJ, Schmechel DE (1987) Neuron specific enolase, a clinically useful marker for neurons and neuroendocrine cells. Ann Rev Neurosci 60:269–295

Martuza RL, Malick A, Markert JM, Ruffner KL, Coen DM (1991) Experimental therapy of human glioma by means of a genetically engineered virus mutant Science 252:854–856

McGeoch DJ, Preston VG, Weller SK, Schaffer PA (1990) Herpes simplex virus. In: O'Brian SJ (ed) Genetic maps. Book 1, Viruses. Cold Spring Harbor, NY, Cold Spring Harbor Laboratory Press, pp 115–120

Miller DG, Adam MA, Miller AD (1990) Gene transfer by retrovirus vectors occurs only in cells that are actively replicating at the time of infection. Mol Cell Biol 10:4239–4242

Roberts RC, DiFiglia M (1989) Short- and long-term survival of large neurons in the excitoxic lesioned rat caudate nucleus: a light and electron microscopic study. Synapse 3:363–371

Rock DL, Beam SL, Mayfield JE (1987) Mapping bovine herpes virus type 1 latency-related RNA in trigeminal ganglia of latently infected rabbits. J Virol 61:3827–3831

Roizman B, Jenkins FJ (1985) Genetic engineering of novel genomes of large DNA viruses. Science 229:1208–1214

Roizman B, Sears AE (1990) Herpes simplex viruses and their replication. In: Fields BN (ed) Virology. New York, Raven Press, pp 1795–1841

Rosenberg WS, Breakefield XO, Isacson O (1991) A potential new method of delivery of an expressed gene to central neurons of the adult rat. Soc Neuroscience Abst 570.7

Schranz P,. Neidhardt H, Schroder CH, Kaerner HC (1989) A viable HSV-1 mutant deleted in two nonessential major glycoproteins. Virology 170:273–276

Sears AE, Hukkanen V, Labow MA, Levine AJ, Roizman B (1991) Expression of the herpes simplex virus 1 alpha transinducing factor (VP16) does not induce reactivation of latent virus or prevent the establishment of latency in mice. J Virol 65:2929–2935

Steiner I, Spivack JG, Lirette RP, Brown SM, MacLean AR, Subak-Sharpe JH, Fraser NW (1989) Herpes simplex virus type 1 latency-associated transcripts are evidently not essential for latent infection. EMBO J 8:505–511

Steiner I, Spivack JG, Deshmane SL, Ace CI, Preston CM, Fraser NW (1990) A herpes simplex virus type 1 mutant containing a nontransinducing Vmw65 protein established latent infection in vivo in the absence of viral replication and reactivates efficiently from explanted trigeminal ganglia. J Virol 64:1630–1638

Stevens JG (1978) Latent characteristics of selected herpes-virus. Adv Cancer Res 26:227–256

Stevens JG, Wagner EK, Devi-Rao GB, Cook ML, Feldman LT (1987) RNA complementary to a herpesvirus alpha gene mRNA is prominent in latently infected neurons. Science 235:1056–1059

Stevens JG (1989) Human herpesvirus: a consideration of the latent state. Microbiol Rev 53:318–332

Turner DL, Cepko CL (1987) A common progenitor for neurons and glia persists in rat retina late in development. Nature 328:131–136

Ugolini G, Kuypers HG, Strick PL (1989) Transneuronal transfer of herpes virus from peripheral nerves to cortex and brainstem. Science 243:89–91

Development of Herpes Simplex Virus as a Gene Transfer Vector for the Central Nervous System

J. C. Glorioso, L. R. Sternberg, W. F. Goins and *D. J. Fink*

Introduction

Herpes simplex virus (HSV) is a large neurotropic virus containing 152 kb of linear dsDNA encoding at least 70 gene products. In natural human infections, the virus replicates primarily in skin or mucosal epithelial cells prior to entering regional axon terminals and retrograde transport of the virus capsid to sensory ganglia, where it may either replicate or establish latency (Roizman and Sears 1990; Fig. 1). Latent virus can be reactivated by a variety of stimuli followed by anterograde transport and productive secondary infection at or near the primary site. In animals models, infection of peripheral sensory ganglia and latency can be achieved through inoculation of the skin (Cook and Stevens 1973), cornea (Seiler and Schwab 1984) or the olfactory bulb after intranasal inoculation (Dobson et al. 1989). Focal infection of specific brain regions can be achieved by stereotactic injection of small volumes of virus into defined brain regions (McFarland et al. 1986; Bak et al. 1977).

Latently infected neurons show an extremely restricted and characteristic pattern of viral gene expression (Hill 1985; Stevens 1989). The only transcriptionally active region of the viral genome maps to a segment of the inverted repeat sequences of the long unique (U_L) region of the viral genome just downstream of the structural gene for infected cell polypeptide 0 (ICPO; Fig. 5). The most abundant transcripts are 1.8 to 2.3 kb in size and have been collectively designated the latency associated transcripts, or LATs (Puga and Notkins 1987; Rock et al. 1987; Spivack and Fraser 1987; Perry et al. 1986; Stevens et al. 1987; Deatly et al. 1987; Wagner et al. 1988). It remains to be proven whether any of the LATS encode a viral protein. Deletions within LAT do not alter the ability of the virus to establish or maintain latency; however, reactivation appears to be impaired among certain LAT mutants (Steiner et al. 1989; Leib et al. 1989a; Ho and Mocarski 1989; Sedarati et al. 1989).

There is growing interest in the use of herpes simplex virus as a vehicle for the delivery of foreign genes to the central nervous system (CNS). HSV probably represents one of very few viral vectors which can deliver and support the expression of foreign genes in adult neurons, since retroviruses require host cell division for integration into the cell genome and neurons are nondividing and post-mitotic (Anderson 1984). By capitalizing on the latent state as a means to effect long-term expression, and by controlling the neurotoxic features of viral

F. H. Gage, Y. Christen (Eds.)
Gene Transfer and Therapy
in the Nervous System

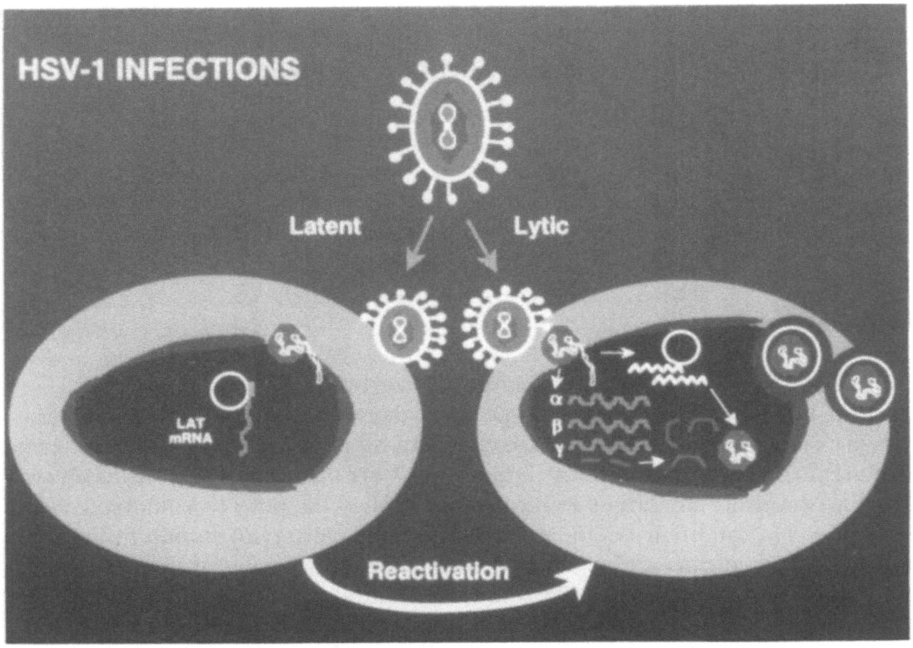

Fig. 1. Life cycle of Herpes simplex virus (HSV). After attachment, fusion, and penetration of a host neuronal cell, the virus capsid migrates to the nucleus where the DNA is released. At this point, the virus may enter either a productive lytic phase or a nonproductive latent state. In the lytic cycle, the viral DNA undergoes replication and viral gene products are expressed in a cascade manner to produce progeny virus. These new viruses are found enveloped in vesicles late in infection and are then released by the infected cell. In the latent state, the virus replicative cycle is interrupted, and the sole transcriptionally active region of the viral genome is limited to the LAT (latency-associated transcript) locus. Latency is a long-term condition which may last the life of the infected host. Reactivation from the latent state may arise after stimulation of the host neuron by such factors as stress, trauma, or certain drugs. When reactivation occurs, the virus becomes transcriptionally active and proceeds through the normal cascade of regulation, leading to the production of infectious viral progeny.

infection, HSV presents an attractive means to develop an effective model system for gene delivery to the brain. To exploit these natural features for gene transfer to brain neurons two fundamental problems must be overcome:

1. a mutant virus must be engineered which does not destroy brain neurons yet is capable of establishing latency in a sufficient number of targeted neurons without the possibility of reactivation; and
2. a means of expressing the transferred gene of interest during latency must be devised, which involves the identification of a latency active promoter/enhancer sequence and appropriate genome location for insertion of the promoter-gene cassette.

Development of a Suitable HSV Vector Backbone

The general strategy for engineering a virus of reduced neuropathogenicity is to delete or interrupt genes which are essential for replicative functions required at different steps in the growth cycle. The propagation of virus mutants lacking essential genes requires a complementing cell line which supplies the essential gene product in trans (De Luca et al. 1985). Alternatively, the removal of one or more nonessential HSV functions might bring about substantial virus attenuation without affecting virus growth in cell culture. Of the over 70 unique HSV genes, at least 30 are know to encode nonessential functions. Viruses with mutations or deletions in the IE genes ICPO and ICP22, for example, establish latency and show a reduced capacity or complete inability to replicate in neurons (Leib et al. 1989b; Sacks and Schaffer 1987; Stoward Stow 1986; Sears at al. 1985; Mc Carthy et al. 1989). Likewise, thymidine kinase (TK) and ribonucleotide reductase (RR) mutant viruses show compromised replication in nerve cells (Coen et al. 1989; Leist et al. 1989; Johnson et al. 1989). As a result, the immediate cytopathology of infected neuronal cells is diminished. Other potential targets include the virion host shutoff (vhs) gene (Kwong and Frankel 1989), a neurovirulence gene (ICP34.5); Chou et al. 1990) and UL46, a modifier of the virus immediate early gene promoter trans-inducing factor (αTIF; Zhang et al. 1991).

Finally, several laboratories have discovered that there are a variety of mutants in the U_s region of the HSV genome which are nonessential but contribute to viral neuropathogenesis (Longnecker and Roizman 1987; Meignir et al. 1988; Weber et al. 1987). Our laboratory devised a rapid insertion mutagenesis strategy in which the bacterial transposable element, Tn5, was used to systematically interrupt the 12 distinct U_s open reading frames (ORFs; Weber et al. 1987). Tn5 insertion results in truncation of the corresponding gene product due to the presence of amber translation termination codons in all three reading frames of the transposon termini. Insertions near the 5' end of the individual U_s genes were selected and incorporated into the viral genome. Except for the US6 gene encoding glycoprotein D, the other 11 U_s genes proved to be nonessential for HSV-1 replication in Vero cells. All of these mutant viruses showed at least a 2 log increase in the LD_{50} following direct intracranial inoculation of mice. A mutation in US3, which encodes a viral protein kinase gene (US3), causes a 4 log increase in LD_{50}, suggesting the potential utility of this mutant virus backbone (Fink et al. 1992). To test this possibility the internal Tn5 sequence of this US3 mutant was replaced with the β-galactosidase *(lacZ)* reporter gene under the control of the HSV glycoprotein C (gC) late gene promoter. The US3::pgC-*lacZ* viral recombinant could then be used to identify neurons in the rat CNS which supported viral replication (late gene expression) and to assess the ability of this mutant virus to spread and replicate in other brain structures.

Studies of US3 Mutant Following Direct Inoculation into Rat Hippocampus

To asses the behavior of the US3::pgC-*lacZ* recombinant virus in rat CNS, 1–10 μl of media containing 10^8 pfu/ml of virus was stereotactically injected unilaterally into the hippocampus and contralateral caudate. A marked blue reaction product indicating β-gal production was seen in the injected hippocampus from one to five days post-inoculation (dpi) and extended throughout the dorsal hippocampus, at least 3 mm in a rostral-caudal direction. Staining was most prominent at two and three days (Fig. 2A and 2B) and disappeared completely by day 10. Few blue neurons were detected in caudate, indicating that the virus replicated poorly in this brain region. In cryostat sections counterstained with neutral red, blue cells are confined to the dentate gyrus of the injected hippocampus. Animals injected with an equivalent number of pfu of wild-type virus (KOS) showed an extensive inflammatory infiltrate at 2 dpi, but no evidence of blue reaction product upon X-gal staining.

The presence of β-gal was confirmed by immunocytochemistry using alternate floating vibratome sections immunostained with a monoclonal antibody specific for β-gal. The immunostaining pattern was comparable to that seen with X-gal (Fig. 2C). This finding demonstrates that the cytochemical reaction product detected results from β-gal gene expression from the virus genome. To determine the identity of the cells supporting β-gal expression, semithin sections prepared from the vibratome sections were examined and the vast majority of cells containing the X-gal reaction product were neurons (Fig. 2D). Electron microscopic examination of ultrathin sections confirmed that the electron-dense X-gal reaction product was localized predominantly in neurons, and

Fig. 2A–B. Transgenic expression of *lacZ* in brain using the US3::pgC-*lacZ* recombinant. At the top of this figure is a schematic representation of the HSV US3::pgC-*lacZ* genome, indicating the location of the pgC-*lacZ* reporter gene cassette. The pgC-*lacZ* reporter cassette, comprised of the HSV late gene glycoprotein C (gC) promoter upstream of the *E. coli* galactosidase *(lacZ)* gene, was cloned as a 4.2 kb BamHI fragment into the BglII sites located within the IS50 elements of Tn5 replacing the internal Tn5 sequences. This Tn5 transposon had previously been inserted into the 5' end of the HSV US3 gene. Since the IS50 elements of Tn5 contain translation termination codons in all three reading frames, insertion of Tn5 into a specific gene results in truncation of that particular gene product. The presence of the Tn5::pgC *lacZ* cassette within the US3 gene locus was confirmed by Southern blot analysis. Five μl of media containing 5×10^5 pfu of virus was stereotactically injected unilaterally into the hippocampus and/or hippocampus and contralateral caudate of 200–250 gm male Sprague Dawley rats. At times ranging from 2 to 300 dpi, rats were perfused with 4% paraformaldehyde, 0.4% glutaraldehyde, and 100 μm virbratome sections were reacted with 0.015% 4-Cl-5-Br-3-indoyl-β-galactopyranoside (X-gal) for 21 hours at 37 °C. A and B, X-gal staining of 100 μm vibratome sections two days after inoculation demonstrates the presence of blue cells unilaterally in the injected hippocampus (A, original magnification × 6; B, original magnification × 50). C, the immunostaining pattern obtained with anti β-gal antibody was comparable to that seen with X-gal staining and was confined to neurons of the dentate gyrus (orignal magnification × 120). D, semithin sections were prepared from the vibratome sections by further fixation overnight in 2% paraformaldehyde-2% glutaraldehyde, followed by dehydration and embedding in epon araldite. X-gal staining of 1 μm sections embedded in plastic shows that the vast majority of blue staining cells had neuronal morphology (original magnification × 1000).

also showed that those cells with X-gal reaction product in the cytoplasm contained viral particles in the nucleus (data not shown). This result is in agreement with the fact that an HSV late gene promoter is responsible for driving *lacZ* expression, correlating reporter gene expression with virus replication. Thus both β-gal activity and viral particles are observed in the same cell. Few enveloped particles were observed in the cytoplasm, however, providing supporting evidence for the limited pathogenesis of this mutant virus.

Stereotactic inoculation of either 5 μl media containing 10^7 pfu/ml wild type HSV-1 (KOS) or a similar dose of a glycoprotein C deletion mutant, gC^- 39, described earlier (Chrisp et al. 1989; Sunstrum et al. 1988); caused death at two to three weeks post-inoculation in two thirds of the injected rats. Surviving animals, sacrificed at three weeks to three months after inoculation, often showed focal encephalomalacia as a result of brain destruction by those viruses of unattenuated pathogenicity. In contrast, the US3::pgC-*lacZ* recombinant showed only limited disruption of the normal neuronal architecture in animals examined four weeks to 20 months after inoculation.

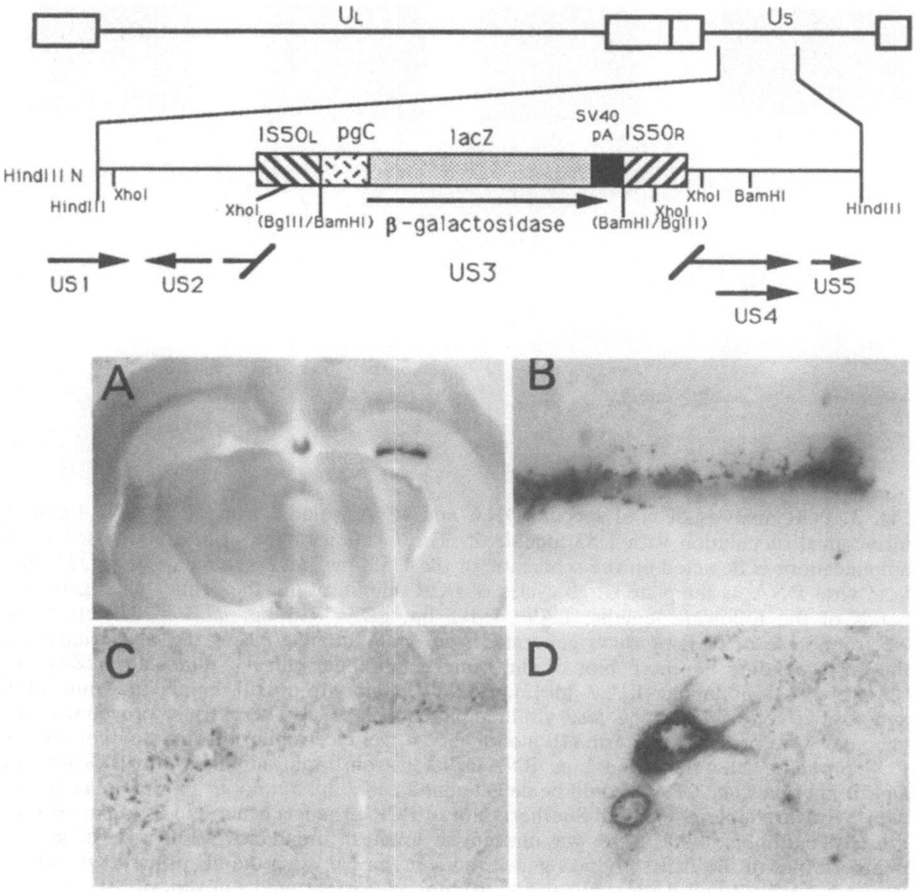

Fig. 2 A–D

To determine whether latent US3::pgC-*lacZ* recombinant viral genomes were present in the brain in the absence of gC promoter driven *lacZ* expression, we performed PCR amplification of DNA and RNA isolated from brains 2 to 300 dpi with primer pairs specific for the HSV glycoprotein B (gB) gene, LAT and for *lacZ*. PCR analysis of DNA isolated from animals sacrificed 2, 7 and 300 dpi demonstrated the presence of the mutant HSV genome at all the time points tested (Fig. 3A). LAT RNA was detected in brains of animals sacrificed 2, 7 and 300 dpi, consistent with expression of LAT during lytic viral replication and from latent viral genomes (Fig. 3C). Both *lacZ* (Fig. 3B) and gB (Fig. 3D) mRNA were only detected in samples from the 2 and 7 dpi brains

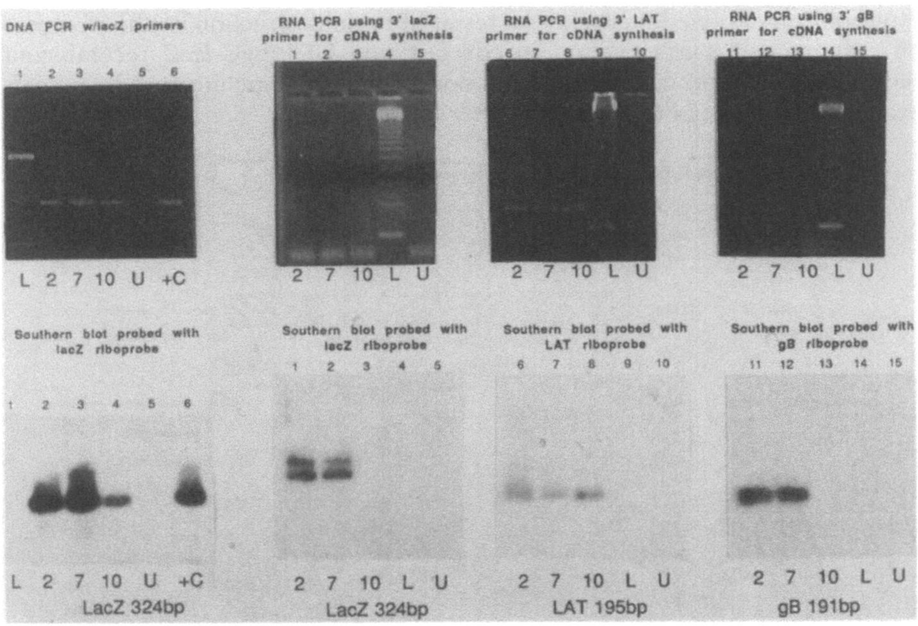

Fig. 3. PCR analysis of viral specific DNA and RNA isolated from RAT CNS following intracranial inoculation with US3::pgC-*lacZ*. The location of the primer pairs used in the amplifications is depicted on the schematic of the US3::pgC-*lacZ* genome. Using US3::pgC-*lacZ* viral DNA as template of 40 cycles of PCR amplification, the primer pairs generated bands of the following lengths: LAT = 195 bp; gB = 191 bp; *lacZ* = 324 bp. a, gel electrophoresis of PCR products generated from brain amplified using the *lacZ* primers and the corresponding Southern blot of the same gel probed with a ^{32}P-labeled *lacZ*-specific riboprobe. 123 bp ladder [L] , 2 dpi [2], 7 dpi [7], 10 month pi [10], uninfected control [U], and positive control US3::pgC-*lacZ* viral template DNA [+C]. These results proved that viral genomes were present long-term (10 months pi). b, gel electrophoresis and Southern blot of PCR products generated from 1 μg RNA isolated from brain amplified 40 cycles with the *lacZ* primer pair. *lacZ* RNA could be detected at 2 and 7 dpi but not at 10 mpi, in agreement with X-gal straining. c, Gel and Southern blot of PCR products generated as in part B using the LAT, primers. LAT RNA was present in greatest abundance at 10 mpi. d, gel and Southern blot of the PCR products generated as in part b using the gB primers. gB mRNA was seen early (2 and 7 dpi), but not at 10 mpi, consistent with the shut-off of lytic HSV genes during latency.

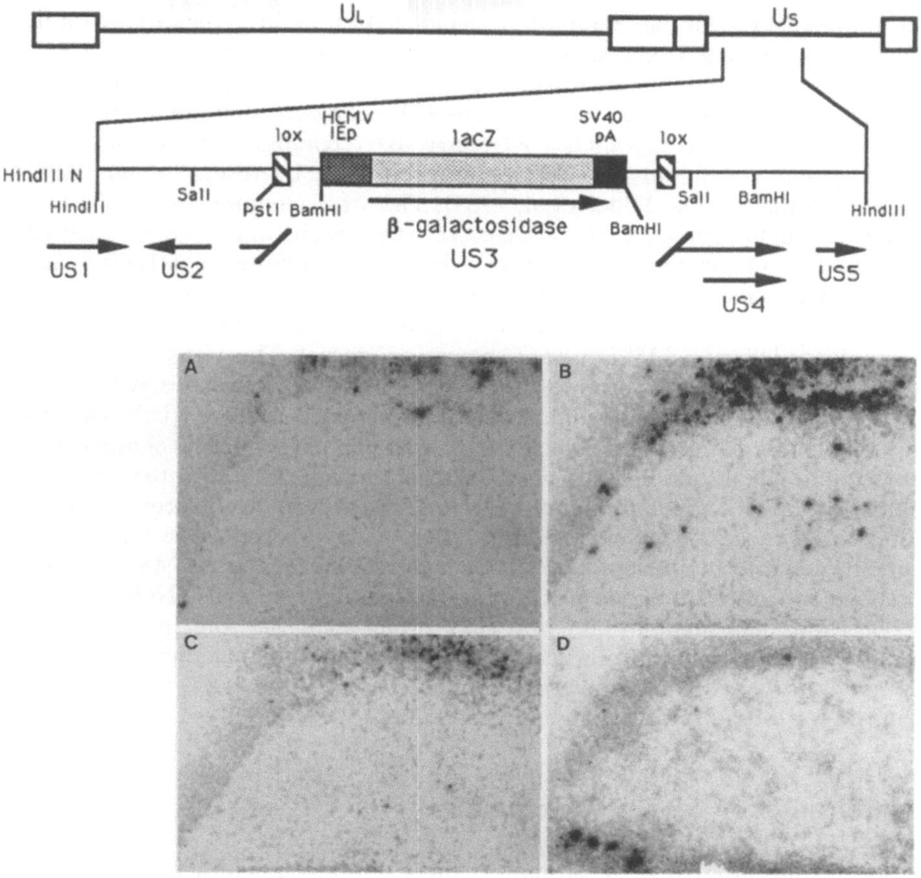

Fig. 4A–D. Expression of *lacZ* and LAT in brain using the US3::HCMV IEp-*lacZ* recombinant virus. The US3::HCMV IEp-*lacZ* mutant was engineered by cre-lox site-specific recombination of a plasmid containing the HCMV IEp-*lacZ* reporter gene cassette into the lox site within the US3 locus of the virus US3::10x3. The location of the HCMV IEp-*lacZ* cassette within the US3 locus is depicted at the top of the figure. Five μl containing 5×10^4 pfu of US3::HCMV IEp-*lacZ* was stereotactially injected into the hippocampus of rats. At 2, 8, 14 and 28 days following inoculation, the animals were sacrificed, and alternate serial cryostat sections were either stained with X-gal and counterstained with neutral red or used for *in situ* hybridizations with riboprobes specific for LAT or ICPO (*A*). *lacZ* expression was detected in cells of the ipsilateral dentate gyrus similar to the US3::pgC IEp-*lacZ* recombinant. In contrast, the US3::HCMV IEp-*lacZ* mutant also displayed *lacZ* expression within neurons of the pyramidal cell layer. *lacZ* expression was transient, with the blue reaction product detected from 2 to 7 dpi. Both the LAT (*B*) and ICPO (*C*) riboprobes gave positive signals at day 2, with the number of LAT positive cells being significantly greater. At 8 dpi, no ICPO positive cells could be detected (data now shown), and the number of LAT positive cells (*D*) was reduced from those seen in two-day animals. All riboprobes were negative on sections from the 14- and 28-day animals.

but not in RNA from the 10-month post-inoculation brains, consistent with the shut-off of lytic gene expression during latency (Stevens 1989).

In an attempt to drive expression of the foreign gene independent of viral replication we reasoned that a strong non-HSV promoter might be sufficient. A virus was constructed in which a P1 phage lox recombination site was introduced into the 5'end of the US3 coding sequence by rescue of an ICP4$^-$ deletion mutant (dl20; Deluca et al. 1985). The virus recombinant is US3$^-$ yet contains a single copy of ICP4. A plasmid containing the human cytomegalovirus (HCMV) major immediate early (IE) gene promoter controlling *lacZ* expression was incorporated into this viral genome using a cell-free site-specific cre recombinase-mediated lox reaction (Sauer et al. 1987); producing the recombinant US3::HCMV IEp-*lacZ* shown in Figure 4. This virus recombinant grew well in cell culture, forming blue plaques upon X-gal agarose overlay, but was highly attenuated upon injection into brain since it lacked US3 and had only one copy of ICP4. Serial sections from brains of inoculated animals were subjected to X-gal staining or *in situ* hybridization with probes for the lytic gene ICPO and LAT. As seen in Figure 4A, the HCMV promoter expressed *lacZ* at 2 dpi; however, *lacZ* gene expression was lost by day 5, despite the fact that the virus is known to be present by PCR detection and that the HCMV promoter has been shown to be active in neurons of transgenic animals (Nelson et al., 1991). These data suggest that a non-HSV promoter in the context of the HSV genome is only active during lytic gene expression and that either the LAT promoter itself or a nonviral promoter inserted into the LAT locus may be required for long-term foreign gene expression.

Identification of the LAT Promoter(s)

To locate the putative LAT promoter, the chloramphenicol acetyltransferase (CAT) reporter gene was introduced at a position 42 bp downstream of the 5' end of the stable LATs. This construct was assayed for CAT expression using transient transfection assays. It has been determined that promoter activity lies within a fragment extending from –598 (# 118867) to +42 (# 119500) relative to the 5' end of the stable LATs (Fig. 5). Co-transfection experiments involv-

Fig. 5. Diagram of the HSV-1 LAT region. The location of the LAT region is shown within the internal repeat segment bracketing the U_L region of the HSV-1 genome at the top of the figure, along with the location of relevant restriction sites within the region, including the numbering of the DraI and BamHI sites according to the nucleotide position from the left-hand end of the prototype genome. Note that the PpuMI site has been changed to a BamHI site in our clone of the LAT region. The region containing LAT promoter sequences is marked as a hatched box. Below is depicted the transcriptional patterns within the LAT region. The RNA transcripts of the HSV genes ICPO and ICP34.5 are shown with the LAT sequences transcribed from the opposite strand. The 8.77 kb polyadenylated transcript is present during productive infection. Its 5' and 3' nucleotide positions are noted. The stable 2 kb LAT transcript, also detected during productive infection, is the major transcript during latency, and the smaller 1.25 kb transcript is a spliced variant found exclusively during latent infection. An expansion of the LAT promoter region displays the two PstI sites as well as the conserved sequence stretching between HSV-1 strains and HSV-2 (vertical lines). Two

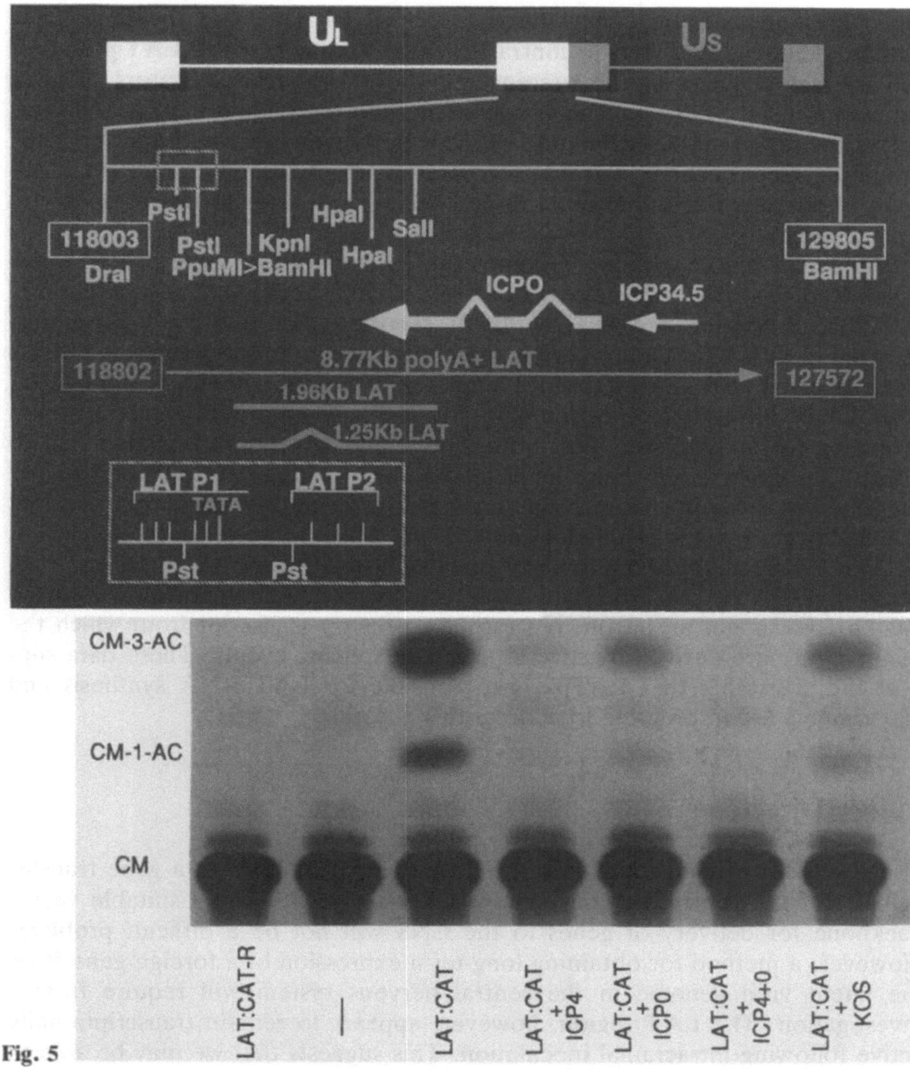

Fig. 5

regions have demonstrated promoter activity, which we label LATp1 and LATp2. LATp1 contains at TATA box, several Sp1 sites and a CRE and is located over 600 bp upstream from the 5' end of the stable 2 kb LAT. LATp2 is located just upstream of the 5' end of the 2 kb stable LAT and does not possess a TATA box. Transient expression analyses using LATp2 is depicted at the bottom of the figure. The LATp2 promoter [Pst I (–598) to bamHI (+42) relative to the 5' end of the 2 kb LAT], juxtaposed to the chloramphenicol acetyltransferase (CAT) reporter gene, was transfected into Vero cells in the absence and presence of equimolar amounts of the HSV gene transactivators ICP4 and ICP0. A construct in which the LATp2 promoter is in the opposite orientation to the CAT gene [LAT-CAT-(R)] was also used. The LAT-CAT construct was transfected into Vero cells, and the cells were infected with wild-type KOS virus (MOI = 5) at 24 hpt. The negative control (–) used a CAT plasmid lacking LATp2 promoter sequences. At 48 hpt, the cells were harvested, and cell lysates were used in CAT assays. Promoter activity is characterized by the presence of the acetylated forms of chloramphenicol (CM-1-AC and CM-3-AC).

ing a plasmid carrying either the ICPO or ICP4 gene and the LAT-CAT plasmid demonstrated that, in contrast to other viral genes, this LAT promoter activity was reduced in the presence of these viral transactivators of gene expression (Fig. 5). This result is consistent with the finding that, as the virus enters the latent state, ICP4 and ICP0 are no longer expressed and the LAT promoter either becomes active or increases in activity, signifying the establishment of latency. For purposes of discussion, we will refer to this promoter as LATp2 (Fig. 5).

While this work was under completion, a number of other laboratories (Devi-Rao et al. 1991; Zwaagstra et al. 1989, 1990; Leib et al. 1991; Melton et al. 1986; Batchelor and O'Hara 1990) reported than the LAT promoter, referred to as LATp1 in this review, was far upstream of the stable LAT RNA and expressed an 8.77 kb transcript initiating 30 bp downstream of the LATp1 TATA box during lytic infection (Fig. 5). The 8.77 kb primary transcript was proposed to be processed post-transcriptionally to produce a stable 2.2 kb intron, the larger LAT transcript detected during latency. Conflicting reports suggested that the upstream promoter was also negatively regulated by ICP4 (Devi-Rav et al. 1991; Batchelor and O'Hara 1990). More recently, Spivack et al. 1991) have mapped multiple 5' ends of the 2.2 kb LAT RNA by PCR assisted cDNA cloning from latently infected ganglia. The data show that the 2.2 kb LAT is not an intron, but rather a primary transcript from which the minor LATs are derived by several alternate splicing events. These data support the possibility that LATp2 is responsible for LAT RNA synthesis and experiments are in progress to address this question.

Future Directions

The results of investigations into the development of HSV as a gene transfer vehicle for the central nervous system suggest that devising a suitable vector backbone for delivery of genes to the CNS will not be a difficult problem. However, a method for obtaining long-term expression of a foreign gene from the latent viral genome in the central nervous system will require further investigation. The LAT region, however, appears to remain transcriptionally active following intracranial inoculation. This suggests that we may be able to exploit the use of the LAT promoters to achieve long-term gene expression in brain. Alternatively, cellular promoters which are known to function constitutively in brain neurons, such as nerve specific enolase (Forss-Petter et al. 1990) or the neurofilament promoter (Julien et al. 1989), may be potentially promising it they are introduced into the LAT region of the genome. These modifications of the LAT regions will be explored as avenues for gene expression from the latent HSV genome in brain.

References

Anderson WF (1984) Prospects for human gene therapy. Science 226:401–409

Bak IJ, Markham CH, Cook ML, Stevens JG (1977) Intraaxonal transport of herpes simplex virus in the rat central nervous system. Brain Res 136:415–429

Batchelor AH, O'Hara P (1990) Regulation and cell type-specific activity of a promoter located upstream of the latency-associated transcript of herpes simplex virus type 1. J Virol 64:3269–3279

Cook ML, Stevens JG (1973) Pathogenesis of herpetic neuritis and ganglionitis in mice: evidence for intra-axonal transport of infection. Infect. Immun 7:272–288

Chou J, Kern ER, Whitley RJ, Roizman B (1990) Mapping of herpex simplex virus-1 neurovirulence to $\gamma_1$34.5, a gene nonessential for growth in culture. Science 250:1262-1266

Chrisp CE, Averill DR, Sunstrum JC, Levine M, Glorioso JC (1989) Characterization of encephalitis in adult mice induced by intracerebral inoculation of herpes simplex virus type 1 (KOS) and comparison with mutants showing decreased virulence. Lab Invest 60:822–830

Coen DM, Kosz-Vnenchak M, Jacobsen JG, Leib DA, Bogard CL, Schaffer PA, Tyler KL, Knipe DM (1989) Thymidine kinase-negative herpes simplex virus mutants establish latency in mouse trigeminal ganglia but do not reactivate. Proc Natl Acad Sci USA 86:4736–4740

Deatly AM, Spivack JG, Lavi E, Fraser NW (1987) RNA from an immediate early region of the type 1 herpes simplex virus genome is present in the trigeminal ganglion of latently infected mice. Proc Natl Acad Sci USA 84:3204–3208

DeLuca NA, McCarthy AM, Schaffer PA (1985) Isolation and characterization of deletion mutants of herpes simplex virus type 1 in the gene encoding immediate-early regulatory protein ICP4. J Virol 56:558–570

Devi-Rao GB, Goodart SA, Hecht LM, Rochford R, Rice MK, Wagner EK (1991) Relationship between polyadenylated and nonpolyadenylated herpes simplex virus type 1 latency-associated transcripts. J Virol 65:2179–2190

Dobson AT, Sedarati F, Devi-Rao G, Flanagan WM, Farrell MJ, Stevens JG, Wagner EK, Feldmann LT (1989) Identification of the latency-associated transcript promoter by expression of rabbit β-globin mRNA in mouse sensory nerve ganglia latently infected with a recombinant herpes simplex virus. J Virol 63:3844–3851

Fink DJ, Sternberg LR, Weber PC, Mata M, Goins WF, Glorioso JC (1992) In vivo expression of β-galactosidase in hippocampal neurons by HSV-mediated gene transfer. Human Gene Ther 3:11–20

Forss-Petter S, Danielson PE, Catsicas S, Battenberg E, Price J, Nerenberg M and Sutcliffe JG (1990) Transgenic mice expressing β-galactosidase in mature neurons under revisions specific enolase promoter control. Neuron 5:187–197

Hill TJ (1985) Herpes simplex virus latency. In The Herpesviruses, B. Roizmann (ed), Plenum Press, New York, pp 175–240

Ho DY, Mocarski ES (1989) herpes simplex virus latent RNA (LAT) is not required for latent infection in the mouse. Proc Natl Acad Sci USA 86:7596–7600

Johnson JG, Leib DA, Goldstein DJ, Bogard CL, Schaffer PA, Weller SK, Coen DM (1989) A herpes simplex virus ribonucleotide reductase deletion mutant is defective for productive acute and reactivatable latent infections of mice and for replication in mouse cells. Virol 173:276–283

Julien PJ, Tretjakoff I, Bendaet L, Peterson A (1987) Expression and assembly of a human neurofilament protein in transgenic mice provide a novel neuronal marking systems. Genes Dev 1:1085–1095

Kwong AD, Frenkel N (1989) The herpes simplex virus virion host shutoff function. J Virol 63:4834–4839

Leib DA, Bogard CL, Kosz-Vnenchak M, Hicks KA, Coen DM, Knipe DM, Schaffer PA (1989) A deletion mutant of the latency-associated transcript of herpes simplex virus type 1 reactivates from the latent state with reduced frequences. J Virol 63:2893–2900

Leib DA, Coen DM, Boargd CL, Hicks KA, Yager DR, Knipe DM, Tyler KL, Schaffer PA (1989) Immediate-early regulatory gene mutants define different stages in the establishment and reactivation of herpes simplex virus latency. J Virol 63:759–768

Leib DA, Nadeau KC, Rundle SA, Schaffer PA (1991) The promoter of the latency-associated transcripts of herpes simplex virus type 1 contains a functional cAMP-response element: role of the latency-associated transcripts and cAMP in reactivation of viral latency. Proc Natl Acad Sci USA 88:48–52

Leist TP, Sandri-Goldin RM, Stevens JG (1989) Latent infections in spinal ganglia with thymidine kinase-deficient herpes simplex virus. J Virol 63:4976–4978

Longnecker R, Roizman B (1987) Clustering of genes dispensible for growth in culture in the S component of the HSV-1 genome. Science 236:573–576

McCarthy AM, McMahan L, Schaffer PA (1989) Herpes simplex virus type 1 ICP27 deletion mutants exhibit altered patterns of transcription and are DNA deficient. J Virol 63:18–27

McFarland DJ, Sikora E, Hotchin J (1986) The production of focal encephalitis in mice by stereotaxic inoculation of virus: anatomical and behavioral effects. J Neurol Sci 72:307–318

Meignier B, Longnecker R, Mavromara-Nazos P, Sears AE, Roizman B (1988) Virulence of and establishment of latency by genetically engineered deletion mutants of herpes simplex virus 1. Virology 162:251–254

Melton DW, McEwan C, KcKie AB, Reid AM (1986) Expression of the mouse HPRT gene: deletional analysis of the promoter region of an X-chromosome linked housekeeping gene. Cell 44:319–328

Nelson J, Baracchini E, Gould S, Ghazal P, Stenberg R, Wiley C (1991) The human cytomegalovirus immediate early promoter targets expression to the brain, salivary gland, pancreas, intestine, and testes of transgenic mice. 16th International Herpesvirus Workshop, p 176

Perry JL, Rixon FJ, Everett RD, Frame MC, McGeoch DJ (1986) Characterization of the IE110 gene of herpes simplex virus type 1. J Gen Virol 67:2365–2380

Puga A, Notkins AL (1987) Continued expression of a poly $(A)^+$ transcript of herpes simplex virus type 1 in trigeminal ganglia of latently infected mice. J Virol 61:1700–1703

Rock DL, Nesbern AB, Ghiasi H, Ong J, Lewis TL, Lokensgard JR, Wechsler S (1987) Detection of latency-related viral RNAs in trigeminal ganglia of rabbits latently infected with herpes simplex type 1. J Virol 61:3820–3826

Roizman B, Sears AE (1990) Herpes simplex viruses and their replication. In Virology, Fields BN, Knipe DM, Chanock RM, Hirsch MS, Melnick JL, Monath TP (eds.) 2nd ed Raven Press, New York, pp 1795–1841

Sacks WR, Schaffer PA (1987) Deletion mutants in the gene encoding the herpes simplex virus type 1 immediate-early protein ICPO exhibit impaired growth in cell culture. J Virol 61:829–839

Sauer BM, Whealy M, Robbins A, Enquist L (1987) Site-specific insertion of DNA into a pseudorabies virus vector. Proc Natl Acad Sci USA 84:9108–9112

Sears AE, Halliburton IW, Meigner B, Silver S, Roizman B (1985) Herpes simplex virus 1 mutant delted in the α 22 gene: growth and gene expression in permissive and restrictive cells and establishment of latency. J Virol 55:338–346

Sedarati F, Izumi KM, Wagner EK, Stevens JG (1989) Herpes simplex virus type 1 latency-associated transcription plays no role in establishment or maintenance of a latent infection in murine sensory neurons. J Virol 63:4455–4458

Seiler M, Schwab ME (1984) Specific retrograde transport of nerve growth factor (NGF) from neocortex to nucleus basilis in the rat. Brain Res 300:33–39

Spivak JG, Fraser NW (1987) Detection of herpes simplex virus type 1 transcripts during latent infection in mice. J Virol 61:3841–3847

Spivack JG, Woods GM, Fraser NW (1991) Identification of a novel latency-specific splice donor signal within the herpes simplex virus type 1 2.0-kilobase latency-associated transcript (LAT): translation inhibition of LAT open reading frames by the intron within the 2.0-kilobase LAT. J Virol 65:6800–6810

Steiner I, Spivak JG, Lirette RP, Brown SM, MacLean AR, Subak-Sharpe JH, Fraser NW (1989) Herpes simplex virus type 1 latency-associated transcripts are evidently not essential for latent infecton. EMBO J 8:505–511

Stevens JG (1989) Human herpesviruses: a consideration of the latent state. Microbiol Rev 53:318–332

Stevens JG, Wagner EK, Devi-Rao GB, Cook ML, Feldman LT (1987) RNA Complementary to a herpes virus α gene mRNA is prominent in latently infected neurons. Science 235:1056–1059

Stow ND, Stow EC (1986) Isolation and characterization of a herpes simplex virus type 1 mutant containing a deletion within the gene encoding the immediate-early peptide Vmw110. J Gen Virol 67:2571–2585

Sunstrum JC, Chrips CE, Averill DR, Levine M, Glorioso JC (1988) Pathogeniticy of glycoprotein C negative mutants of herpes simplex virus type 1 for the mouse central nervous system. Virus Res 11:17–32

Wagner EK, Devi-Rao G, Feldmann LT, Dobson AT, Zhang Y-F, Flanagan WM, Stevens JG (1988) Physical characterization of the herpes simplex virus latency-associated transcript in neurons. J Virol 62:1194–1202

Weber PC, Levine M, Glorioso JC (1987) Rapid identification of nonessential genes of herpes simplex virus type 1 by Tn5 mutagenesis. Science 236:576–579

Zhang Y, Sirko DA, McKnight JLC (1991) Role of herpes simplex virus type 1 UL46 and UL47 in alpha TIF-mediated transcriptional induction: characterization of three deletion mutants. J Virol 65:829–841

Zwaagstra JC, Ghiasi H, Nesburn AB, Wechsler SL (1989) In vitro promoter activity associated with the latency-associated transcript gene of herpes simplex virus type 1. J Gen Virol 70:2163–2169

Zwaagstra JC, Ghiasi H, Slanina SM, Nesburn AG, Wheathley SC, Lillycrop K, Wood J, Latchman DS, Patel K, Wechsler SL (1990) Activity of herpes simplex virus type 1 latency-associated transcript (LAT) promoter in neuron-derived cells: evidence for neuron specificity and for a large LAT transcripts. J Virol 64:5019–5028

Dystrophin mRNA Processing in the Canine Homologue of Duchenne Muscular Dystrophy: An Authentic Model for Gene Therapy

N. J. H. Sharp, J. N. Kornegay, S. D. van Camp, M. H. Herbstreith, S. L. Secore, S. Kettle, M. J. Dykstra, C. D. Constantinou, A. D. Roses, and *R. J. Bartlett*

Duchenne muscular dystrophy (DMD) is a fatal, X-linked, recessive disease of humans that afflicts 1 in 3500 live-born males (Koenig et al. 1988). Approximately two thirds of DMD patients carry detectable deletions in the gene encoding dystrophin (Walker et al. 1989; Koenig et al. 1987), which is a muscle cytoskeletal protein (Koenig et al. 1988). Dystrophin protein (Hoffmann et al. 1988) and transcript (Oronzi Scott et al. 1988) are either absent or severely deficient in DMD patients (Koenig et al. 1988).

Genetic therapy strategies in experimental animals are extremely valuable when directed against the authentic disease. Although the mdx mouse model of dystrophin deficiency involves a mutation in the same gene (Sicinski et al. 1989), the mouse does not develop weakness or dystrophy in the same manner as boys with DMD (Koenig et al. 1988; Hoffman et al. 1988; Partridge 1991). In fact, the mouse has small fibers and is very strong, almost a super-mouse when the muscle is viewed on dissection (Partridge 1991).

Golden retriever muscular dystrophy (GRMD) is a homologous condition in which affected dogs lack dystrophin transcript and protein (Cooper et al. 1988). The phenotypes of dogs with GRMD and DMD patients are strikingly similar (Kornegay et al. 1988; Valentine et al. 1988). We have defined the genetic defect in GRMD and initiated specific genetic therapy experiments based on these data.

We have detected a truncated dystrophin transcript in affected dogs using the polymerase chain reaction (PCR; Saiki et al. 1988). Our data indicate that this truncation is due to precise deletion of the seventh exon from the dystrophin transcript, due to a mutation in either the sixth or seventh intron (Sharp et al., 1992).

To characterize the defect in GRMD, we have amplified single-stranded dystrophin cDNA prepared from skeletal muscle of normal and affected dogs. All primers for the reverse transcriptase and PCR reactions (Fig. 1) were based on the human dystrophin cDNA sequence (Koenig et al. 1988). Products amplified by PCR were visualized after Southern blotting and subsequent hybridization with human skeletal muscle dystrophin cDNA (Koenig et al.1987).

Using PCR primer pair F1/R2 (exons 1–8), a canine dystrophin cDNA of expected size (1 kb) was amplified from the muscle of three normal dogs (Fig. 2 A, lanes 1, 2, 8). Using the same primer pair, a slightly smaller dystro-

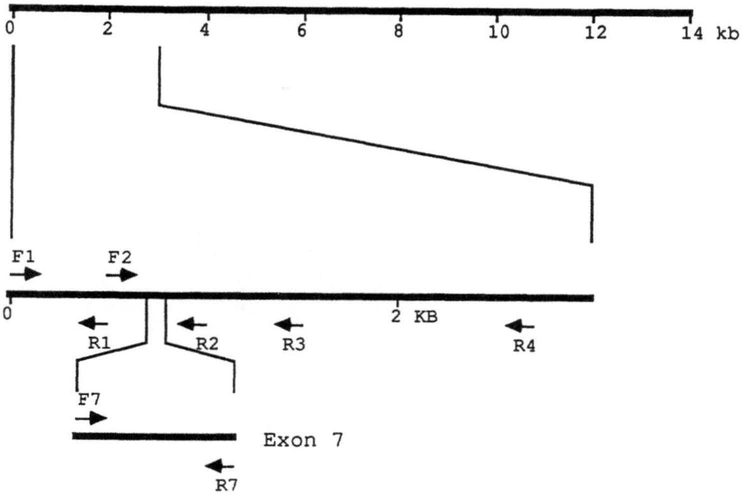

Primer Positions and Sequence

		5' 3'	
F1	(1-24)	GGGATTCCCTCACTTTCCCCCTAC	EXON 1
F2	(529-565)	CTTGGTTTGATTTGGAATATAATCCTCCACTGGCAG	EXON 5
F7	(738-766)	GCCAGACCTGCTTGACTGGAATAGTGTG	EXON 7
R1	(500-479)	CAGTACTTCCAATATTCACTAG	EXON 5
R2	(990-969)	GTCACTTTAGGTGGCCTTGGC	EXON 8
R3	(1518-1476)	GCTACCCTGAGGCATTCCCATCTTGAATTTAGGAGATTCATC	EXON 11
R4	(2709-2679)	GCGATGATGTTGTTCTGATACTCCAGCCAG	EXON 20
R7	(857-829)	CTTCAGGATCGAGTAGTTTCTCTATGCC	EXON 7

Fig. 1. Oligonucleotide primers used for reverse transcription and cDNA PCR. The primers are shown relative to their position in the dystrophin cDNA (Koenig et al. 1988). Forward (F) and reverse (R) primers are depicted by arrows; scale on the dystrophin cDNA is in kilobases. Oligo-nucleotide primers were prepared using DuPont DNA synthesizers (Coder 300 and Generator).

phin cDNA (approximately 0.88 kb)) was amplified from each of five different dogs with GRMD (Fig. 2A, lanes 3–7). Using a new reverse transcription product and PCR primer pair F1/R3 (exons 1–11), the expected 1.5 kb dystrophin cDNA was also approximately 120 nucleotides smaller in GRMD dogs compared to normal dogs (data not shown). Similar results were obtained using primer pair F1/R4 (exons 1–20) and amplification of the 2.7 kb fragment of the 5′ end of the canine dystrophin cDNA (data not shown). In contrast, expected products of equal size were amplified from the first 0.5 kb of normal and mutant canine dystrophin cDNA using primer pair F1/R1 (exons 1–5, data not shown). These four experiments together suggested a deletion in the GRMD

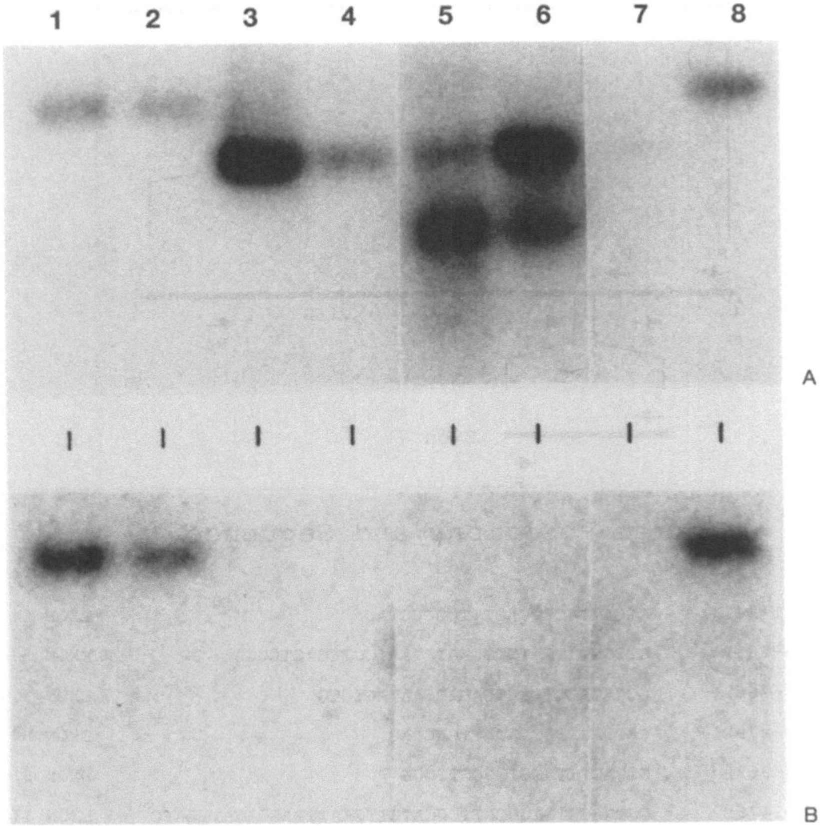

Fig. 2A, B. PCR amplification of normal and GRMD transcripts. Amplified products of normal and GRMD mRNA were separated on agarose gels, blotted and hybridized (Feener et al. 1989). *A* was probed with 9-7 cloned fragment of human dystrophin cDNA (nucleotides 1–1500, relative to published cDNA sequence; Koenig et al. 1988). Lanes 1, 2, 8 are samples from normal dogs; lanes 3–7 are samples from GRMD dogs. *B* was probed with end-labelled (Sambrook et al. 1989) oligo-nucleotide F7 located within exon seven of the human cDNA. Lanes are as per Figure 2A. Size standards are indicated in base pairs. Total RNA (4 ug. per sample) from skeletal muscle (limb) was prepared with guanidinium hydrochloride (Chirgwin et al. 1979) and reverse transcribed into first strand cDNA with an oligo-nucleotide primer R2, and Superscript reverse transcriptase (BRL). Primer pair F1/R2 was used to amplify first strand cDNA using 20 cycles, standard buffers and 2.5 units of Taq polymerase (Cetus). Annealing was at 37 °C for 10 minutes for the first five cycles, then a 55 °C for the next 15 cycles. Extention was at 72 °C for 2 to 3 minutes, and denaturation was at 94 °C for 1 minute.

transcript between primers R1 and R2 (corresponding to nucleotides 565 and 969, exons 5–8, of the human dystrophin cDNA).

Product generated by the first three experiments (primer pairs F1/R2, F1/R3, F1/R4) was used as template in a second, asymmetric PCR amplication. Single stranded cDNA produced by PCR primer pair F2/R2 (exons 5–8) was sequenced to determine the basis for the truncated product from GRMD muscle (Fig. 3A). The sequence obtained from the cDNA of normal canine muscle corresponded very closely to that of human dystrophin (Koenig et al. 1988; Bartlett et al. 1990). In particular, exon seven was present in the product from normal canine muscle and comprised 119 nucleotides. The sequence of the truncated cDNA from GRMD dogs was equivalent up to and including the last nucleotide of exon six. The next nucleotide in the GRMD sequence corresponded to the first nucleotide of exon eight from the normal dog. Exon seven was precisely deleted and there was no apparent inclusion of sequence that might correspond to that from the intervening sequences between exons six and eight (Fig. 3B). This finding was confirmed in three normal and three affected dogs and in the amplification products generated from three different cDNAs reverse transcribed from GRMD mRNA. The size of the deleted exon corresponded very closely to the difference in PCR products generated in normal and GRMD dogs. Assuming that intron-exon junctions in dogs and humans are

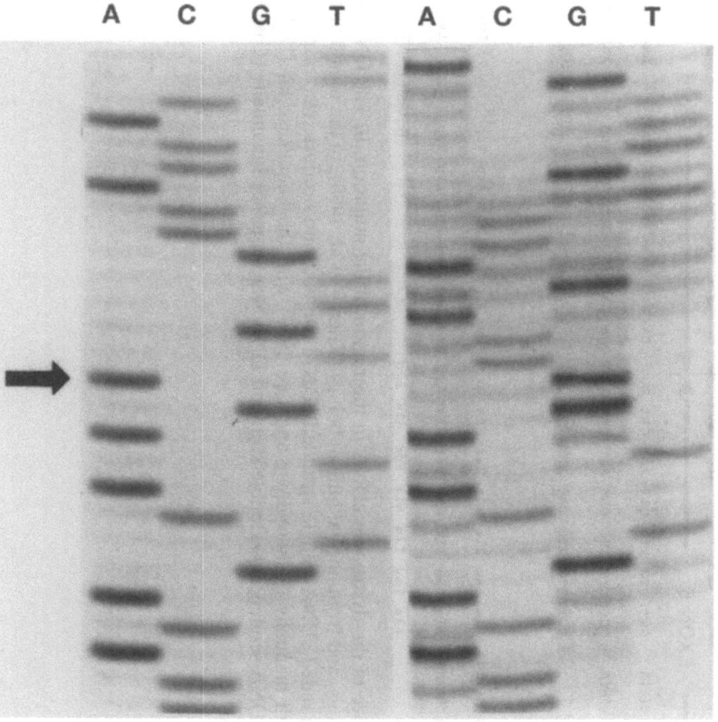

Fig. 3. Sequence of the dystrophin cDNA in the region of the mutation. **a.** Canine dystrophin sequence from normal (4 lanes on right) and GRMD (4 lanes on left) dogs in the region of the mutation. The sequence representing exon seven is deleted from the GRMD dog (arrow) when compared to the normal dog. This change has been observed in three normal and three GRMD dogs, and in different amplification products from the same GRMD mRNA.

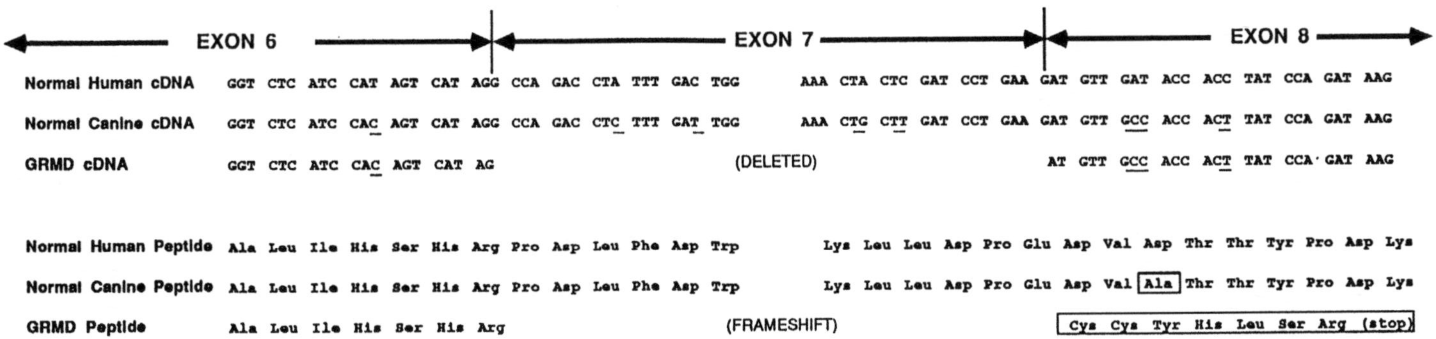

Fig. 3. Sequence of the dystrophin cDNA in the region of the mutation. **b.** The DNA sequences and predicted amino acid sequences from normal and GRMD dogs, and from normal human beings (Koenig et al. 1988), in the region of the mutation. Primer pair F2/R2 was used to generate single and double stranded cDNA. Single stranded cDNA amplification products were separated by electrophoresis through an agarose gel, purified and sequenced both manually (Mihilovilovic and Lee 1989) with the Sequenase kit (USB) and by an automated sequence analysis system (DuPont Genesis 2000 DNA sequencer). Double stranded DNA was sequenced manually (Casanova et al. 1990).

identical, the skipping of exon seven would change the reading frame of exon eight. This would predict a premature termination codon in the eighth codon of the new reading frame (Fig. 3B).

To confirm that the truncated PCR product of GRMD muscle was caused by the skipping of exon seven, oligo-nucleotide (F7) was end-labelled (Sambrook et al. 1989) and hybridized to the PCR products amplified in the first three experiments. In each case, only products amplified from normal canine muscle produced a signal (Fig. 2B). No major rearrangements have been detected by screening genomic blots of normal and GRMD dogs using the entire human dystrophin cDNA (Bartlett et al. 1990). A labelled PCR product (Schowalter and Sommer 1989) specific for exon seven (primer pair F7/R7) gave identical restriction fragments when hybridized to Southern blots of genomic DNA from normal and GRMD dogs (Fig. 4). This indicates that exon seven is present in the genome of GRMD dogs and that there is no obvious deletion or insertion in intron six or seven.

Dystrophin transcript had not been detected by previously published Northern blot analysis of skeletal muscle RNA from GRMD dogs (Cooper et al. 1988). We prepared a Northern blot using total RNA from normal and GRMD canine skeletal muscle. This was hybridized with probe representing the first 1500 nucleotides of human dystrophin cDNA. A 14 kb transcript was present in normal dogs but dystrophin mRNA was not detected in the GRMD samples (Fig. 5). The same results were observed using exon seven specific primers (data not shown). Using PCR technology, we have identified that low levels of dystrophin transcript are, in fact, present in GRMD muscle. This discrepancy

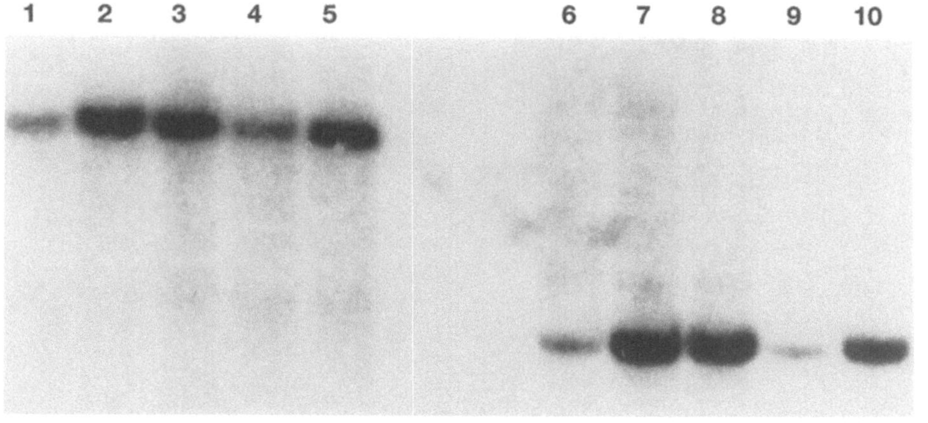

Fig. 4. Southern blot analysis of genomic DNA from normal and GRMD dogs. DNA samples in lanes are as follows: 1, 6, GRMD male; 2, 3, 7 and 8, carrier females; 4,9, normal male; 5, 10 normal female. Lanes 1–5 are Eco R1 digested DNA and lanes 6–10 are HindIII digested DNA. Intensity differences reflect X-chromosome dosage effects. Samples were separated by electrophoresis through a 1% agarose gel, transferred to Gene Screen Plus membranes (Dupont) and hybridized to a PCR generated probe (Schowalter and Sommer 1989) representing exon seven made using pair F7/R7.

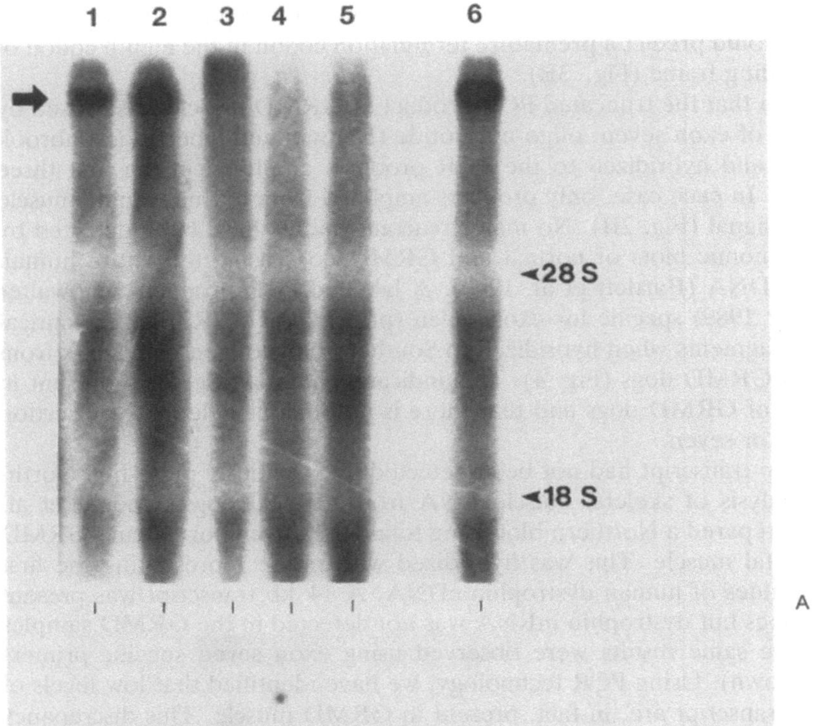

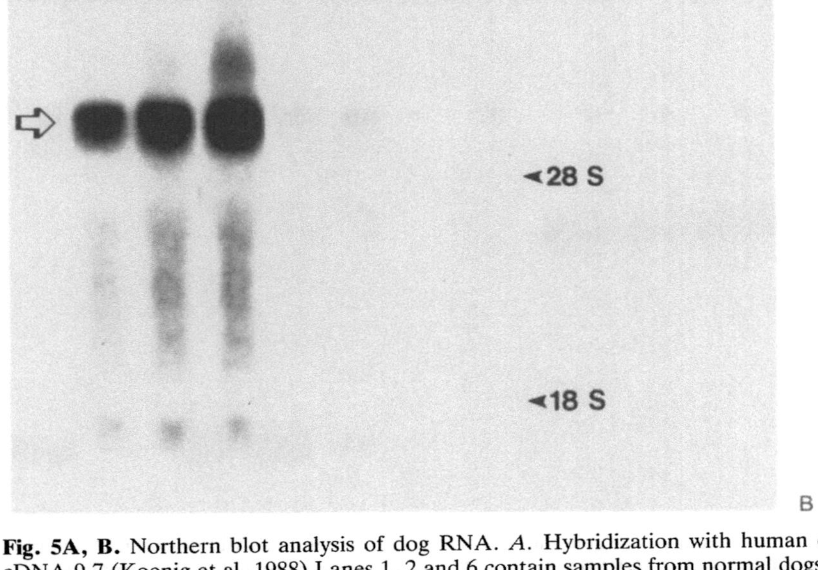

Fig. 5A, B. Northern blot analysis of dog RNA. *A*. Hybridization with human dystrophin cDNA 9-7 (Koenig et al. 1988) Lanes 1, 2 and 6 contain samples from normal dogs and lanes 3–5 from GRMD dogs. Lanes 1–3 are from adult dogs and lanes 4–6 are from neonatal dogs. A 14 kb transcript is evident in lanes 1, 2 and 6 but not in lanes 3–5. These same results were obtained using the exon seven specific probe (data not shown). *B*. Evidence that RNA is not degraded is shown by labelled adult human myosin heavy chain cDNA (Saez and Leinward 1986) probe on the same blot. The weak signal in lanes 4–6 is due to the limited cross-

presumably reflects the greater sensitivity of PCR to detect low abundance sequences (Saiki et al. 1988).

Alternative splicing of canine dystrophin mRNA is a potential explanation for our findings. In humans, the primary dystrophin transcript in brain contains a different first exon than muscle and the 3′ end of the dystrophin transcript may also be alternatively spliced (Feener et al. 1989). However, amplification of the human muscle dystrophin segment from nucleotides 141 to 1207 did not reveal alternative splicing of exons from this region (Feener et al. 1989). Amplification of a novel alternative transcript is an unlikely explantation for the truncated PCR product seen in GRMD dogs for two reasons. No such transcript was seen in the normal siblings examined in this study. In addition, alternative splicing of exon seven predicts a frameshift mutation (Fig. 3B). This would induce a premature termination codon and therefore probably explains the lack of dystrophin previously shown in GRMD (Kornegay et al. 1990).

Splice site mutations which result in processing errors of dystrophin pre-mRNA have not been described in either DMD or the dystrophin-deficient mdx mouse. A nonsense mutation at position 3714 has been described in one DMD patient (Bulman et al. 1991). In the mdx mouse, a point mutation in nucleotide 3185 creates a premature termination codon (Sicinski et al. 1989), and the levels of both dystrophin transcript and protein are reduced (Hoffman et al. 1987; Chamberlain et al. 1988). Nonsense mutations in GRMD dogs and mdx mice are the probable direct cause of the reduced mRNA levels in their muscles. Nonsense mutations also reduce mRNA levels in human beta-thalassemia, whereas missense mutations do not (At weh et al. 1988; Baserga and Benz 1988).

Many of the human beta-thalassemias are due to RNA processing errors (Mitchell et al. 1986). The more common of these mutations prevent splicing of the beta-globin pre-mRNA at a particular site and activate formerly silent cryptic sites (Treisman et al. 1983). This usually results in deletion of part of an exon or inclusion of intronic sequence in the mutated transcript (Treisman et al. 1983). Such an event was not detected in GRMD; rather, a precise skipping of an exon occurred. Exon skipping is uncommon in beta-thalassemia (Mitchell et al. 1986), but is the main mechanism underlying the processing errors seen in the Nagase analbuminemic rat (Salaby and Shafritz 1990), and in human osteogenesis imperfecta (Bonadio et al. 1990), Ehlers-Danlos syndrome type VII (Weil et al. 1989), porphyria (Grandchamp et al. 1989), phenylketonuria (Dilella et al. 1986), and retinoblastoma (Horowitz et al.1989). Exon skipping results from inactivation of an RNA splice site and is usually caused by a single base mutation within the consensus splice sequence (Salaby and Shafritz 1990; Bonadio et al. 1990; Weil et al. 1989; Srandchamp et al. 1989; Dihella et al.

---

hybridization between the probe for adult MHC and the neonatal myosin. Further evidence was the intensity of the 28S ribosomal RNA bands in ethidium stained gels and the use of the same RNA samples for PCR analysis. Samples (20 ug) of total RNA from normal and GRMD limb skeletal muscle were isolated (Chirgwin et al. 1979), separated by electrophoresis through a 1% agarose/formaldehyde gel, transferred to Hybond membranes (Amersham), and hybridized with cDNA probes as described(Sambrook et al. 1989).

1986; Horowitz et al. 1989). A mutation such as this is also the explanation for the precise skipping of exon seven during mRNA processing that we have identified in GRMD (Sharp et al. 1992) and might also explain the defect found in some DMD patients who lack genomic deletions Walker et al. 1989, 1992; Koenig et al. 1987). We are currently assessing DMD patients who lack deletions using RNA/PCR to determine if similar errors in RNA processing are present.

Definition of the genetic defect of GRMD should also facilitate use of this model in therapy trials that may have application to DMD. The GRMD model should be particularly appropriate for these studies as, unlike the mdx mouse, which is essentially clinically normal, affected dogs have progressive clinical disease than can be objectively assessed during treatment. Concerning potential therapies for DMD, considerable attention has been focused on gene (Assadi et al.1991; Klamput et al. 1991) and myoblast transfer therapies (Partridge 1991). Although we plan to begin minigene trials in the GRMD model soon, our research thus far has focused on myoblast transfer therapy. To address certain basic questions, we have conducted initial experiments in normal dogs injured with the muscle-specific toxin, notexin (Harris et al. 1975). Notexin is a phospholipase that is a constituent of Australian tiger snake venom and selectively necrotizes mature myofibers while sparing nerves, blood vessels, and myoblasts. Our studies have shown that notexin produces a consistent pattern of muscle necrosis and regeneration within 21 days in dogs (Dykstra et al. 1990; Sharp et al., unpublished or in press, see attached acceptance letter). Muscles to be transplanted are irradiated before notexin injury to lessen this host-derived regeneration response (Gulati 1987), and autologous cells are transplanted to avoid immunorejection. Mixed cultures thought to contain predominantly myoblasts are used and cells are labelled with fluorescein microspheres before transplantation (Alameddine et al. 1989). Staged and concomitant notexin injury and myoblast transplantation have been studied (Kornegay et al. 1991). Results obtained after concomitant injury/transplantation have been most promosing. Large clusters of microsphere labelled cells were seen seven days after injury/transplantation, with apparent fusion after 21 days. In contrast, relatively few microsphere-labelled cells were seen at either time point when transplants were done several days after the injury. Concomitant notexin injury and transplantation apparently facilitates myoblast implantation, although the responsible mechanisms are not yet clear.

Most of the microsphere-labelled cells noted in our studies remained in distinct clusters apart from preexisting host muscle. Transplanted cells apparently fused with each other principally, rather than with preexisting myoblasts. However, the microsphere labelling system does not allow for critical assessment of chimeric fiber formation. We plan to use chromosomal differences between male and female cells to make this distinction. By using molecular probes for the Y chromosome, male cells can be distinguished from those of females (Beilharz et al. 1991). A canine-specific Y chromosome probe has been developed (Beilharz et al. 1991), and should also be useful in studying female donor/male host pairs using our notexin injury model. The Y chromosome probe should be very useful in studying myoblast transplantation in the GRMD model. By breeding affected males to obligate carriers, homozygous affected females can be obtained. Donor myoblasts from normal male siblings can thus be distinguished in female GRMD dogs. Thus the system will follow several parameters: donor myoblasts in the transplant, dystrophin production, and functional recovery. Therapy with human minigene constructs has been initiated using this model.

Acknowledgments. The authors wish to thank L. M. Kunkel for human dystrophin cDNA clones and L. Leinward for human adult myosin heavy chain probe. Support for this research was from the Association Francaise contre les Myopathies (AMF to RJB and JNK); Muscular Dystrophy Association (JNK); March of Dimes (JNK); National Institutes of Neurological Disease and Stokes, NS26630 (RJB.ADR); National Institute on Aging LEAD Award AG07922 (ADR); and the Joseph and Kathleen Bryan Alzheimer's Disease Research Center (ADR).

References

Acsadi G, Dickson G, Love DR, Jani A, Walsh FS, Gurusinghe A, Wolfe JA, Davies KE (1991) Human dystrophin expression in mdx mice after intramuscular injection of DNA constructs. Nature 352:815–818

Alameddine HS, Dehaupas M, Fardeau M (1989) Regeneration of skeletal muscle fibers from autologous satellite cells multiplied in vitro: An experimental model for testing cultured cell myogenicity. Muscle Nerve 12:544–555

Atweh GF, Brickner HE, Zhu XX, Kazazian HH Jr, Forget BG (1988) New amber mutation in a Beta-Thalassemia gene with nonmeasurable levels of mutant mRNA in vivo. J Clin Invest 82:557–561

Bartlett RJ, Sharp NJH, Secore SL, Hung WY, Kornegay NJ, Roses AD (1990) The canine and human DYS genes are highly conserved. J Neurol Sci 98 (Suppl):165

Baserga SJ, Benz EJ Jr (1988) Nonsense mutations in the human beta-globin gene affect mRNA metabolism. Proc Natl Acad Sci USA 85:2056–2060

Beilharz M, Garrett K, Fan Y, Fletcher S, Lareu R, Maley M, Vague M, Harvey A, Grounds M (1991) Genetic probes for tracking muscle precursor cell in vivo: Technical Aspects. In: Kakulas BA, Howell JMc, Roses AD (eds) Animal models for duchenne muscular dystrophy and genetic manipulation workshop, 7–9 August, 1991, Raven Press, in press.

Bonadio J, Ramirez F, Barr M (1990) An intron mutation in the human alpha1 (I) collagen gene alters the efficiency of pre- mRNA splicing and is associated with osteogenesis imperfecta type II. J Biol Chem 265:2262–2268

Bulman DE, Gangopadhyay SB, Bebchuk KG, Worton RG, Ray PN (1991) Point mutation in the human dystrophin gene: Identification through western blot analysis. Genomics 10:457–460

Casanova JL, Pannetier C, Jaulin C, Kourilsky P (1990) Optimal conditions for directly sequencing double-stranded PCR products with Sequenase. Nucleic Acids Research 18:4028

Chamberlain JS, Pearlman JA, Muzny DM, Gibbs RA, Ranier JE, Reeves AA, Caskey CT (1988) Expression of the murine Duchenne muscular dystrophy gene in muscle and brain. Science 239:1416–1418

Chirgwin JM, Przybyla AE, MacDonald RJ, Rutter WJ (1979). Isolation of biologically active ribonucleic acid from sources enriches in ribonuclease. Biochemistry 18:5294–5299

Cooper BJ, Winand NJ, Stedman H, Valentine BA, Hoffmann EP, Kunkel LM, Orzoni Scott M, Fischbeck KH, Kornegay JN, Avery RJ, Williams JR, Schmickel RD, Sylvester JE (1988) The homologue of the Duchenne locus is defective in X-linked muscular dystrophy of dogs. Nature 334:154–156

Dilella AG, Marvit J,Lidksy AS, Guttler F, Woo SLC (1986) Tight linkage between a splicing mutation and a specific DNA haplotype in phenylketonuria. Nature 332:799–803

Dykstra MJ, Sharp NJH, Bogan DJ, Kornegay JN (1990) Notexin-induced muscle injury in dogs. J Neurol Sci 98 (suppl):127

Feener CA, Koenig M, Kunkel LM (1989) Alternative splicing of human dystrophin mRNA generates isoforms at the carboxy terminus. Nature 338:509–511

Grandchamp B, Picat C, de Rooij F, Beaumont C, Wilson P, Deybach JC, Nordman Y (1989) A point mutation G> A in exon 12 of the porphobilinogen deaminase gene results in exon skipping and is responsible for acute intermittant porphydria. Nucleic Acids Res 17:6637–6649

Gulati AK (1987) The effect of X-irradiation on skeletal muscle regeneration in the adult rat. J Neurol Sci 78:111–120

Hoffmann EP, Brown RH Jr, Kunkel LM (1987) Dystrophin: the protein product of the Duchenne muscular dystrophy locus. Cell 51:919–928

Harris JB, Johnson MA, Karlsson E (1975) Pathological responses of rat skeletal muscle to a single subcutaneous injection of a toxin from the venom of the Australian tiger snake. Clin Exp Pharm Phys 2:383–404

Hoffmann EP, Fischbeck KH, Brown RH, Johnson ML, Medori R, Loike JD, Harris JB, Waterson R, Brooke M, Specht L, Kupsky W, Chamberlain J, Caskey CT, Shapiro F, Kunkel LM (1988) Dystrophin characterization in muscle biopsies from Duchenne and Becker muscular dystrophy patients. New Engl J Med 31:1363–1368

Horowitz JM, Yandell DW, Park S-H, Canning S, Whyte P, Buchkovich K, Harlow E, Weinberg RA, Dryja TP (1989) Point mutation and activation of the retinoblastoma antioncogene. Science 243:937–940

Klamut HJ, Bosnoyan LO, Worton RG, Ray PN (1991) Expression of a human dystrophin minigene in myogenic cell cultures. In: Kakulas BA, Howell JMc, Rosses AD (eds) Animal models for Duchenne muscular dystrophy and genetic manipulation workshop. 7–9 August 1991, Raven Press, in press

Koenig M, Hoffmann EP, Bertelson CJ, Monaco AP, Feener C, Kunkel LM (1987) Complete cloning of the Duchenne muscular dystrophy (DMD) cDNA and preliminary genomic organization of the DMD gene in normal and affected individuals. Cell 50:509–517

Koenig M, Monaco AP, Kunkel LM (1988) The complete sequence of dystrophin predicts a rod-shaped cytoskeletal protein. Cell 53:219–228

Kornegay JN, Tuler SM, Miller DM, Levesque DC (1988) Muscular dystrophy in a litter of golden retriever dogs. Muscle Nerve 11:1056–1064

Kornegay JN, Sharp NJH, Bartlett RB, Van Camp SD, Burt CT, Hung WY, Kwock L, Roses AD (1990) Golden retriever muscular dystrophy: Monitoring for success. In: Eastwood AB, Griggs RC, Karpati G (eds) Advances in Experimental Medicine and Biology, Plenum Press, New York 280:267–272

Kornegay JN, Prattis SM, Bogan DJ, Sharp NJH, Bartlett RJ, Alameddine HS, Dykstra MJ (1991) Results of myoblast transplantation in a canine model of muscle injury in: Kakulas BA, Howell JMc, Roses AD (eds) Animal models for Duchenne muscular dystrophy and genetic manipulation workshop. 7–9 August 1991, Raven Press, in press

Mihilovilovic M, Lee JE (1989) An efficient method for sequencing PCR amplified DNA. Biotechniques 7:14–16

Mitchell PJ, Urlaub G, Chasin L (1986) Spontaneous splicing mutations at the dihydrofolate reductase locus in Chinese hamster ovary cells. Molec Cell Biol 6:1926–1935

Oronzi Scott M, Sylvester JE, Heiman-Patterson T, Shi Y-J, Fieles W, Stedman H, Burghes A, Ray P, Worton R, Fischbeck KH (1988) Duchenne muscular dystrophy gene expression in normal and diseased human muscle. Science 239:1418–1420

Partridge TA (1991) Invited Review: Myoblast transfer: A possible therapy for inherited myopathies. Muscle Nerve 15:197–212

Saez L, Leinward LA (1986) Characterization of diverse forms of myosin heavy chain expressed in adult human skeletal muscle. Nucleic Acids Res 14:2951–2969

Saiki RK, Gelfand DH, Stoffel S, Scharf SJ, Higuchi R, Horn GT, Mullis KB, Ehrlich HA (1988) Primer-directed enzymatic amplification of DNA with a thermostable DNA polymerase. Science 239:487–491

Salaby F, and Shafritz DA (1990) Exon skipping during splicing of albumin mRNA precursors in Nagase analbuminemic rats. Proc Natl Acad Sci USA 87:2652–2656

Sambrook J, Fritsch EF, Maniatis T (1989) Molecular cloning: a laboratory manual. 2nd edition. Cold Springs Harbour Laboratory, Cold Springs Harbor, NY

Schowalter DB, Sommer SS (1989) The generation of radiolabelled DNA and RNA probes with polymerase chain reaction. Annal Biochem 177:90–4

Sharp NJH, Kornegay JN, Van Camp SD, Herbstreith M, Secore S, Hung WY, Dykstra MJ, Roses AD, Bartlett RJ (1992) Exon skipping during dystrophin mRNA processing in a canine model of Duchenne muscular dystrophy. Genomics 13:115–121

Sharp NJH, Kornegay JN, Bartlett RJ, Hung WY, Dykstra MJ. Notexin – induced muscle injury in the dog. J Neurol Sci (in press)

Sicinski P, Geng Y, Ryder-Cook AS, Barnard EA, Darlison MG, Barnard PJ (1989) The molecular basis of muscular dystrophy in the mdx mouse: a point mutation. Science 244:1578–1580

Treisman R, Orkin SH, Maniatis T (1983) Specific transcription and RNA splicing defects in five cloned Beta-Thalassemia genes. Nature 302:591–596

Valentine BA, Cooper BJ, deLahunta A, O'Quinn R, Blue JT (1988) Canine X-linked muscular dystrophy. An animal model of Duchenne muscular dystrophy: clinical studies. J Neurol Sci 88:69–81

Walker AP, Bartlett RJ, Laing NG, Hung WY, Yamaoka LH, Secore SL, Holsti M, Speer MC, Mechler F, Denton M, Siddique T, Pericak-Vance MA, Roses AD (1989) A partial Taq I map of the Duchenne muscular dystrophy gene: use of small cDNA fragments in deletion and RFLP analysis. Am J Human Genet 45:A166

Walker AP, Laing NG, Yamada T, Chandler DC, Kakulas B, Bartlett RB (1992) A Taq1 map of the dystrophin gene useful for deletion and carrier status analysis. J Med Genet 29:14–19

Weil D, D'Alessio M, Ramirez F, de Wet W, Cole WC, Chan D, Bateman JF (1989) A base substitution in the exon of a collagen gene causes alternative splicing and generates a structurally abnormal peptide in a patient with Ehlers-Danlos syndrome type VII. EMBO J 8:1705–1710

Gene Transfer of a Murine Dystrophin Minigene Construct

P. R. Clemens, C. C. Lee, and *C. T. Caskey*

Summary

Successful gene correction of dystrophin-deficient muscle cells depends on both the availability of a functional dystrophin cDNA construct and an understanding of the conditions required for efficient gene transfer and subsequent stable expression. A 14 kilobase mouse dystrophin cDNA was cloned by splicing together two overlapping partial cDNA constructs at a unique *Kpn*I restriction site. The clone was full-length, as determined by selective sequencing and restriction mapping. Western blot analysis using an anti-dystrophin antibody demonstrated that electroporation of the construct into COS cells yielded transient expression of a protein of relative molecular mass 430 000. Recombinant dystrophin expression was localized predominantly to the COS cell membrane by immunohistochemical detection using an anti-dystrophin antibody.

To study the expression of the recombinant dystrophin construct *in vivo*, a modified vector based on the muscle creatine kinase promoter with the E1 and E2 enhancers was developed to drive the dystrophin cDNA. Microinjection of this construct into a mouse ovum resulted in a transgenic mouse with a single copy integration of the dystrophin cDNA. Dystrophin expression in *mdx* progeny which carry the transgene was demonstrated at the mRNA level by a reverse transcriptase-polymerase chain reaction method and at the protein level by Western blot assay. The transgene dystrophin protein, identical in size to native muscle dystrophin, is expressed in skeletal and cardiac muscle.

Introduction

Duchenne muscular dystrophy (DMD) is an X-linked disease with an incidence of 1 in 3500 male births. The major phenotypic feature is progressive degeneration of skeletal muscle that is caused by the absence or abnormality of a muscle cytoskeletal protein, dystrophin. The identification of both the dystrophin gene and the dystrophin protein has revolutionized the diagnosis of this disorder. By gene mutation analysis alone, which requires only a blood sample, 65% of affected males can be detected. Furthermore, anti-dystrophin antibodies are used in Western blot analysis to identify the dystrophin protein and, if present, to assess the quantity and molecular mass of the protein. Immunohistochemistry using anti-dystrophin antibodies provides cellular localization of dystrophin in a tissue section (reviewed in Chamberlain and Caskey 1990). However, because of a high spontaneous mutation rate causing DMD, accurate diagnosis of affected males with subsequent identification of carrier female relatives cannot eradicate this devastating disorder. Therefore, several approaches toward therapy are underway.

The dystrophin gene is large, encompassing 2.3×10^6 base pairs (bp) on the short arm of the X chromosome (den Dunnen et al. 1989). The 14 kilobase (kb) mRNA codes for a 427 kilodalton(kd) protein, dystrophin (Koenig et al. 1987; 1988). Dystrophin is a cytoskeletal protein and has structural homology to two other cytoskeletal proteins found in muscle, α-actinin and spectrin (Koenig et al. 1988). The four protein structure domains that are predicted from the cDNA sequence are:

1. an amino-terminal domain with homology to the actin binding domain of α-actinin,
2. a repeat domain thought to form a rod-like structure with homology to spectrin,
3. a cysteine-rich domain with homology to α-actinin, and
4. a carboxy-terminal domain with homology to a recently isolated autosomal transcript called DMDL (Love et al. 1989).

Dystrophin protein is detected in skeletal, cardiac, and smooth muscle and in brain (Chamberlain et al. 1988; Hoffman et al. 1988b). Immunohistochemical studies using anti-dystrophin antibodies localize dystrophin to the sarcolemma of skeletal muscle fibers (Arahata et al. 1988; 1989; Hoffman et al. 1988a; Zubrzycka-Gaarn et al. 1988). Dystrophin forms a tight complex with at least four membrane glycoproteins (Campbell and Kahl 1989; Ohlendieck et al. 1991). The concentration of one of these dystrophin-associated glycoproteins has been shown to be greatly reduced in DMD patients (Ervasti et al. 1990).

Here, we review our studies in which we isolated a full-length, functional murine dystrophin cDNA and demonstrated its expression in a cell culture expression system and in transgenic mice. These studies provide a foundation for human gene transfer experiments in DMD.

Construction of a full-length dystrophin cDNA

From poly (A)+ mouse skeletal muscle RNA, two λ ZAP cDNA libraries corresponding to the 5′ and 3′ ends of the dystrophin cDNA were constructed using oligonucleotide primers beginning at nucleotide position 7,858 and sequence adjacent to the poly(A) sequence, respectively. Three probes, corresponding to the 5′, central, and 3′ regions of the dystrophin cDNA, were used to screen these two libraries to isolate a pair of overlapping clones. The 5′ clone had a 7.8 kb insert and the 3′ clone had an 8.0 kb insert. The clones were excised from the λ ZAP vector and spliced at a unique *Kpn*I restriction site in the region of overlap between the two clones. This cDNA, determined to be full-length by selected sequencing and restriction mapping, was cloned into the

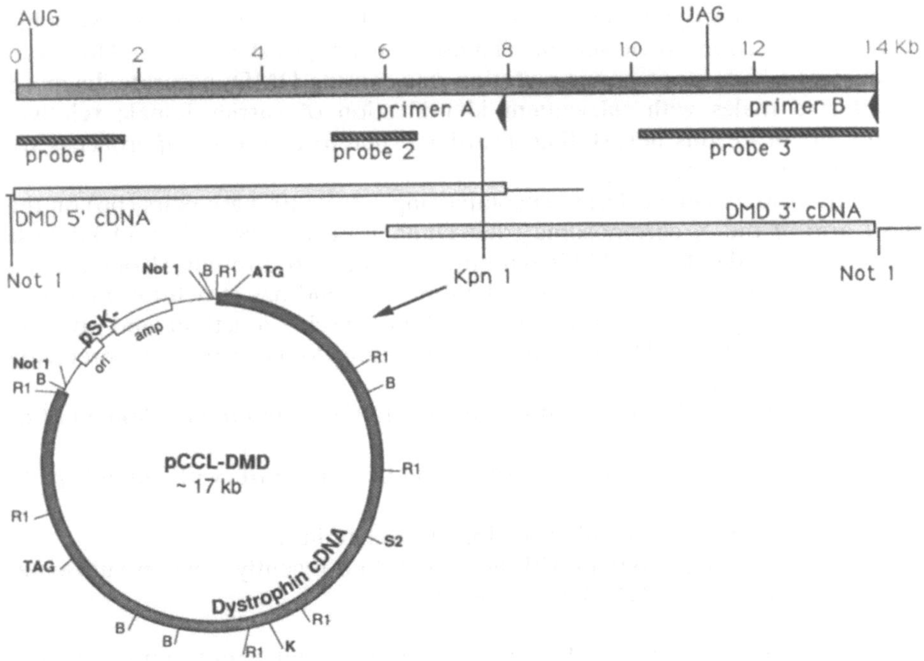

Fig. 1. Cloning of the dystrophin cDNA. The dystrophin mRNA is shown schematically indicating the locations of the start and stop codons and the locations of primer A (5′-GTTCTAGCTTCTTGATTGCTGG-3′) and primer B (5′-TTGTAACATAACTGCG-TGCT-3′), used to generate the 5′ and 3′ cDNA libraries, respectively. Probes 1 (nucleotides 7-1834 of the dystrophin cDNA) and 2 (nucleotides 5079-6546 of the dystrophin cDNA) were used to screen the 5′ library and probes 2 and 3 (nucleotides 10381-poly A of the dystrophin cDNA) were used to screen the 3′ library. The overlapping 5′ and 3′ clones and their shared, unique *Kpn*I site used to splice the two clones together are shown schematically. The resulting full-length dystrophin cDNA clone was subcloned into pSK- to generate pCCKL-DMD. Restriction sites within pCCL-DMD are *Bam*HI (B), *Eco*RI (R1), *Sac*II (S2) and *KpnI* (K). (From Lee et al. 1991; used by permission.)

*Not*I site of the bluescript vector pSK- to generate pCCL-DMD (Fig. 1; Lee et al. 1991).

Expression of the dystrophin cDNA in COS cells

To investigate the expression of the recombinant dystrophin gene, the full-length cDNA was cloned into the *Sma*I site of the mammalian SV40 expression vector pSVL to generate pSVL-DMD. This vector was introduced into COS (African green monkey kidney) cells by electroporation. Western blot analysis of COS cell extracts 72 hours after electroporation showed expression of dystrophin (Fig. 2). The detected protein had a relative molecular mass of approximately 430 kd, the same size as dystrophin detected in an extract of normal

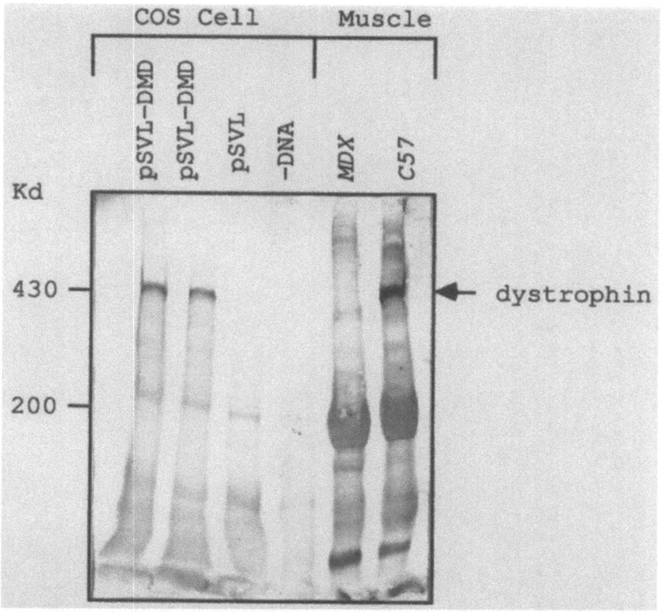

Fig. 2. Western blot demonstrating dystrophin expression in COS cells after electroporation of a recombinant dystrophin cDNA construct. The first four lanes are COS cell extracts prepared 72 hours after transfection of pSVL-DMD (15 μg and 7.5 μg), pSVL (15 μg) and no DNA. Electroporation of 10^7 COS cells was performed at 960 μF and 200 V in 0.8 ml phosphate buffered saline. Lanes 5 and 6 are extracts of hind limb muscle from *mdx* and C57 mice, respectively. Dystrophin was detected by the alkaline phosphatase procedure using an anti-dystrophin polyclonal antibody that was preadsorbed with COS cell extract followed by a biotinylated anti-sheep antibody. (From Lee et al. 1991; used by permission).

C57 mouse muscle, and was not seen in COS cells undergoing electroporation without DNA or with pSVL that did not contain the insert. To immunolocalize the recombinant dystrophin within transfected COS cells, the dystrophin cDNA was then cloned into the pcDL-SRα296 vector that contains a more active promoter, SRα (Takebe et al. 1988). The construct composed of this vector with the dystrophin cDNA insert, called pSRα-DMD, was electroporated into COS cells. Immunocytochemistry done on the transfected cells after acetone fixation demonstrated dystrophin expression that appeared to localize predominantly to the periphery of the cell (Fig. 3). Immunostaining was performed using a primary polyclonal anti-dystrophin antibody raised in sheep (Hoffman et al. 1987) that was preadsorbed with COS cell extract, followed by an anti-sheep antibody conjugated with fluorescein isothiocyanate. Localization of dystrophin to the membrane in COS cells, which do not have endogenous dystrophin, suggests that targeting of dystrophin within the cell may be directed by the protein itself (Lee et al. 1991).

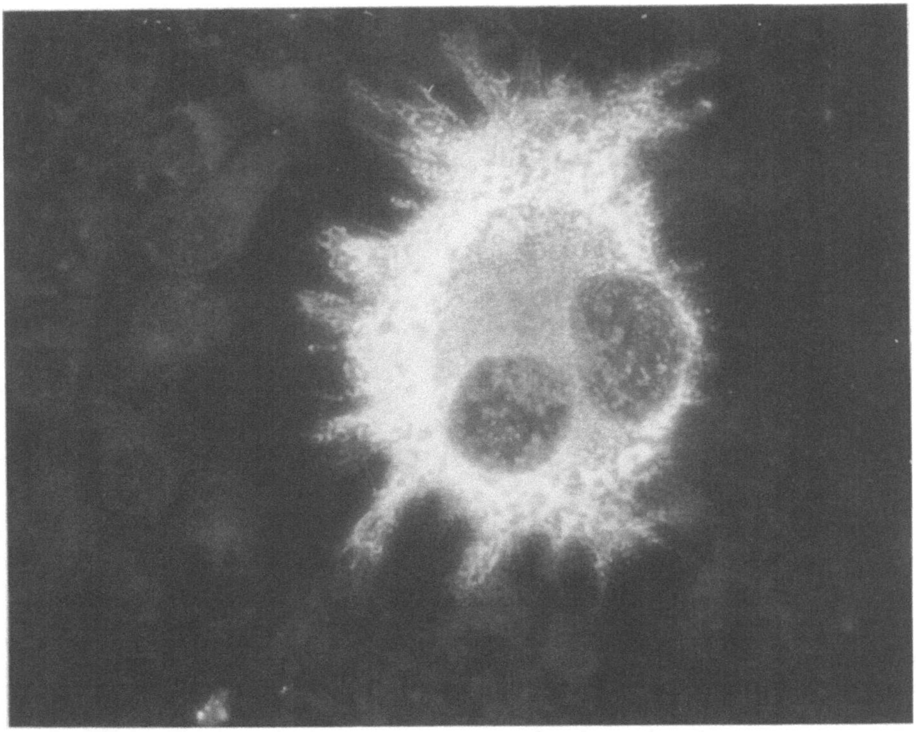

Fig. 3. Localization of recombinant dystrophin in COS cells. Transfection was performed by electroporation of 10^7 COS cells with 15 μg pSRα-DMD. Cells were fixed in acetone 72 hours after electroporation and blocked with 1% bovine serum albumin in phosphate buffered saline. A 2-hour incubation with an anti-dystrophin polyclonal antibody that was preadsorbed with COS cell extract was followed by a 1-hour incubation with an anti-sheep antibody conjugated with fluorescein isothiocyanate to immunolocalize dystrophin within the transfected cells. Note that immunofluorescence appears most intense at the cell rim. (From Lee et al. 1991; used by permission).

Preparation of vectors for transgenic studies

The pCCLMCK-II vector was constructed by cloning the 4.5 kb *Pst*I/*Sal*I fragment from the vector pCK4800CAT (Sternberg et al. 1988) into the *Sal*I/*Pst*I site of the vector pSK$^-$. A 0.6 kb *Bgl*II fragment containing the E2 enhancer from the plasmid pSTA was then cloned into the *Bgl*II site at – 1800 bp from the MCK transcription start site. Electroporation experiments into the mouse C2 muscle cell line (Yaffe and Saxel 1977) under a variety of conditions showed approximately 5- to 10-fold higher CAT activity with pCLLMCK-II than with pCK4800CAT, suggesting that pCCLMCK-II has a stronger promoter. The recombinant dystrophin cDNA was cloned into pCCLMCK-II to generate a vector called pCCLMCK-DMD (Fig. 4). The prokaryotic vector sequences were excised from the MCK promoter, the enhancers and the dys-

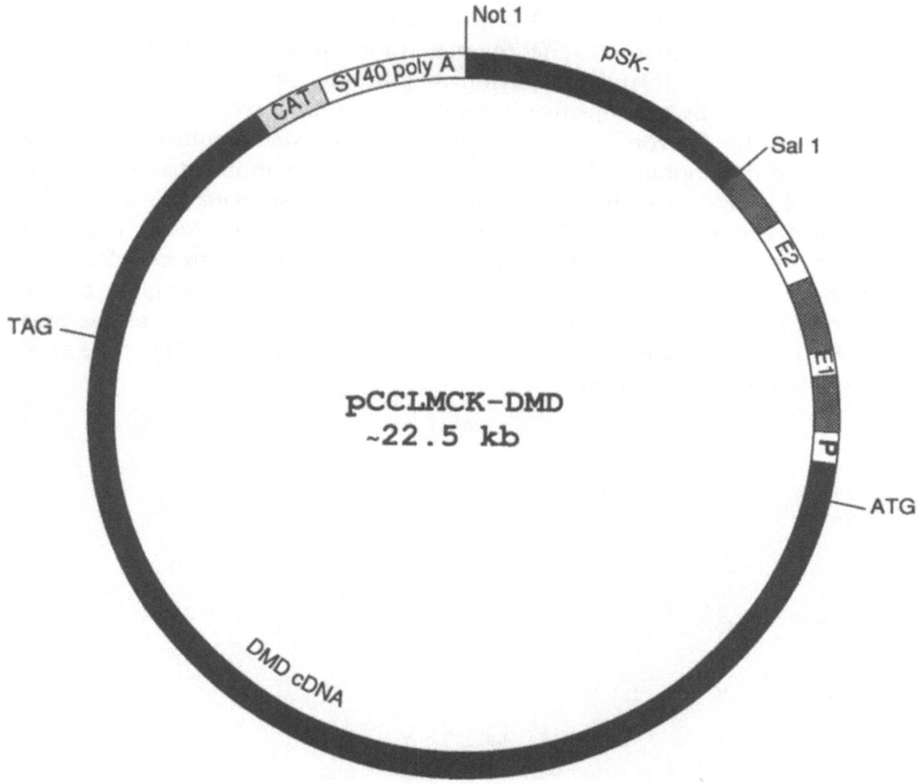

Fig. 4. Construction of pCCLMCK-DMD. The dystrophin cDNA construct is driven by the MCK promoter and the E1 and E2 enhancers. After excision of the prokaryotic vector sequences, this construct was used for embryo microinjection.

trophin cDNA before use for embryo microinjection to produce a transgenic mouse (Lee et al., submitted for publication).

Recombinant dystrophin mRNA and protein expression in transgenic mice

Microinjection of the recombinant dystrophin cDNA driven by the MCK promoter and E1 and E2 enhancers into the pronucleus of a normal FVB mouse embryo produced a male transgenic mouse with a single copy integration of the transgene. This mouse was bred with *mdx* females to generate male progeny expressing the transgene on an *mdx* genetic background.

To determine which tissues expressed dystrophin message, first strand cDNA was generated by reverse transcription of RNA extracted from different tissue types of the transgene-carrying *mdx* mouse. Then, the products of the first strand cDNA reaction were amplified by the polymerase chain reaction (PCR) using primers specific for the CAT gene sequences which were part of the dystrophin cDNA construct. The reverse transcriptase-PCR reactions were electrophoresed, transferred to a nylon membrane, and hybridized with the CAT gene. This is a very sensitive method to determine which tissues express message from the dystrophin cDNA construct and avoids confusion from the lower level of dystrophin mRNA present in some *mdx* mouse tissues (Chamberlain et al. 1988). By this assay, mRNA expression from the dystrophin cDNA construct is seen in skeletal and cardiac muscle, to a lesser degree in spleen, and in trace amounts in kidney, liver, brain and smooth muscle.

Consistent with the findings of dystrophin message levels, dystrophin protein expression of the expected size was demonstrated in skeletal and cardiac muscle by Western blot analysis using anti-dystrophin antibodies (Fig. 5); Lee et al., submitted for publication).

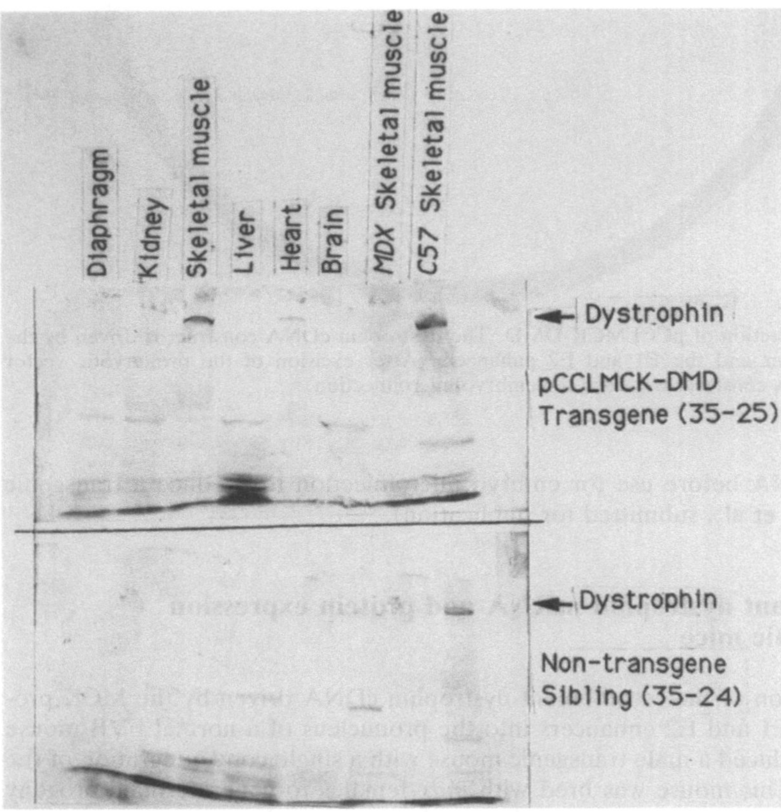

Fig. 5. Western blot analysis of recombinant dystrophin protein expression in the *mdx* mouse with a single copy integration of the transgene. Dystrophin expression is detected in skeletal and cardiac muscle tissue. Extracts of C57 and *mdx* mouse skeletal muscle show presence and absence of dystrophin, respectively. No dystrophin was detected in tissues from a *mdx* sibling that did not carry the transgene.

Conclusions and future studies

These studies demonstrate conclusively that a recombinant murine dystrophin cDNA was isolated and, after microinjection into a mouse embryo, was expressed in the expected pattern of tissues at both the protein level and the mRNA level. The success of these experiments in an animal model provides a firm foundation for human gene transfer studies toward the goal of gene therapy for DMD patients. Ideally, methods will be developed for systemic administration of a human dystrophin gene construct that will be capable of targeting the appropriate tissues. At the present time, known replication-deficient viral vectors cannot carry an insert as large as the full-length dystrophin cDNA. Therefore, research must be directed at developing novel vectors capable of holding a larger insert, at reducing the size of the dystrophin cDNA without significantly affecting function, and at developing new methods of gene transfer. Perhaps a combination of these approaches will yield a practical and effective method of gene therapy for DMD patients.

Acknowledgments. C.T.C. is an investigator of the Howard Hughes Medical Institute. This work was supported by a Neuromuscular Disease Genetics Research Grant from the Muscular Dystrophy Association.

References

Arahata K, Ishiura S, Ishiguro T et al (1988) Immunostaining of skeletal and cardiac muscle surface membrane with antibody against Duchenne muscular dystrophy peptide. Nature 333:861–863

Arahata K, Hoffman EP, Kunkel LM et al (1989) Dystrophin diagnosis: comparison of dystrophin abnormalities by immunofluorescence and immunoblot analyses. Proc Natl Acad Sci USA 86:7154–7158

Campbell KP, Kahl SD (1989) Association of dystrophin and an integral membrane glycoprotein. Nature 338:259–262

Chamberlain JS, Caskey CT (1990) Duchenne muscular dystrophy. In: Appel S (ed) Current neurology. Mosby-Year Book, Inc, Chicago, pp 65–103

Chamberlain JS, Pearlman JA, Muzny DM et al. (1988) Expression of the murine Duchenne muscular dystrophy gene in muscle and brain. Science 239:1416–1418

den Dunnen JT, Grootscholten PM, Bakker E et al. (1989) Topography of the Duchenne muscular dystrophy (DMD) gene: FIGE and cDNA analysis of 194 cases reveals 115 deletions and 13 duplications. Am J Human Genet 45:835–847

Ervasti JM, Ohlendieck K, Kahl SD, Gaver MG, Campbell KP (1990) Deficiency of a glycoprotein component of the dystrophin complex in dystrophic muscle. Nature 345:315–319

Hoffman EP, Brown RH Jr, Kunkel LM (1987) Dystrophin: The protein product of the Duchenne muscular dystrophy locus. Cell 51:919–928

Hoffman EP, Fischbeck KH, Brown RH et al. (1988a) Characterization of dystrophin in muscle-biopsy specimens from patients with Duchenne's or Becker's muscular dystrophy. New Engl J Med 318:1363–1368

Hoffman EP, Hudecki MS, Rosenberg PA, Pollina CM, Kunkel LM (1988b) Cell and fiber-type distribution of dystrophin. Neuron 1:411–420

Koenig M, Hoffman EP, Bertelson CJ, Monaco AP, Feener C, Kunkel LM (1987) Complete cloning of the Duchenne muscular dystrophy (DMD) cDNA and preliminary genomic organization of the DMD gene in normal and affected individuals. Cell 50:509–517

Koenig M, Monaco AP, Kunkel LM (1988) The complete sequence of dystrophin predicts a rod-shaped cytoskeletal protein. Cell 53:219–228
Lee CC, Pearlman JA, Chamberlain JS, Caskey CT (1991) Expression of recombinant dystrophin and its localization to the cell membrane. Nature 349:334–336
Love DR, Hill DF, Dickson G et al (1989) An autosomal transcript in skeletal muscle with homology to dystrophin. Nature 339:55–58
Ohlendieck K, Ervasti JM, Snook JB, Campbell KP (1991) Dystrophin-glycoprotein complex is highly enriched in isolated skeletal muscle sarcolemma. J Cell Biol 112:135–148
Sternberg EA, Spizz G, Perry WM, Vizard D, Weil T, Olson EN (1988) Identification of upstream and intragenic regulatory elements that confer cell-type-restricted and differentiation-specific expression on the muscle creatine kinase gene. Mol Cell Biol 8:2896–2909
Takebe Y, Seiki M, Fujisawa J et al (1988) SR alpha promoter: an efficient and versatile mammalian cDNA expression system composed of the simian virus 40 early promoter and the R-U5 segment of human T-cell leukemia virus type 1 long terminal repeat. Mol Cell Biol 8:466–472
Yaffe D, Saxel O (1977) Serial passaging and differentiation of myogenic cells isolated from dystrophic mouse muscle. Nature 270:725–727
Zubrzycka-Gaarn EE, Bulman DE, Karpati G et al. (1988) The Duchenne muscular dystrophy gene product is localized in sarcolemma of human skeletal muscle. Nature 333:466–469

Gene Therapy in Animal Models of Neurological Disorders

J. Mallet, P. Horellou, C. Lundberg, P. Brundin, P. Kalen, P. Colin, and *A. Björklund*

Summary

The relative importance of synaptic versus paracrine dopamine release for the expression of functional effects following intrastriatal grafting has not, been established. In the present study, we grafted cell lines expressing form I of human tyrosine hydroxylase after infection with a recombinant retrovirus and selection in tyrosine-free-medium to dopamine (DA) denervated striatum. Our goal was to analyse the extent to which extracellular DA levels could be restored and to study the effect of diffuse release of DA on motor impairment in a rat model of Parkinson's disease. In a petri dish, modified fibroblast cells (NIH 3T3) release DOPA constitutively whereas modified endocrine cells (RIN) store and release dopamine in a regulated way. Interestingly, in denervated striatum, the DOPA produced by grafts of modified fibroblast cells was efficiently converted into dopamine by the host striatal tissue. In the grafted striatum, both fibroblast and endocrine cells restored normal levels of extracellular fluid DA. Release of dopamine from fibroblasts was notably unaffected by high concentrations of potassium, in agreement with the *in vitro* properties of the grafted cells. The intrastriatal grafts of modified cells partially reversed the apomorphine-induced but not the amphetamine-induced motor asymmetry. We discuss the implications of these results for the treatment of Parkinson, disease in man.

Introduction

Over the last decade intracerebral transplantation has emerged as a new tool for the study of repair and recovery of function after experimental brain damage. Initial studies concerned Parkinson's disease (PD), a neurodegenerative disorder characterized by a loss of dopaminergic cells in the substantia nigra (Ehringer and Hornykiewicz, 1960).

Animal models of PD can be obtained by experimental lesion of the dopaminergic nigrostiatal pathway following injection of a specific neurotoxin such as 6-hydroxydopamine (6-OHDA; Ungerstedt and Arbuthnott, 1970). The unilateral lesion in rats following such injections results in spontaneous turning behaviour and following the by administration of dopaminergic agonists. In such rats, striatal grafts of foetal mesancephalic neurones have shown their capacity to reinnervate the host striatum, restore striatal dopamine synthesis and release, and compensate for some of the cardinal sensory-motor impairments (Brundin et al. 1987; Schmidt et al. 1982, 1983; Björklund and Stenevi et al. 1979; Björklund et al. 1980; Dunnett et al. 1981, 1988; Zetterström et al. 1986; Strecker et al. 1987; Herman et al. 1985, 1988; Mandel et al. 1990). The efficiency of this method has led to clinical trials of grafts of foetal ventral mesencephalon in patients with PD (Lindvall et al. 1990; Freed et al. 1990). The results of these initial trials have been encouraging, but it is clear that the current transplantation techniques require further development to reach the stage when transplantation can be offered as a routine therapy for PD. One of the major problems with clinicalneural grafting is the source of donor tissue. First, the use of human foetal brain tissue for transplantation is a much debated issue from an ethical standpoint. In several countries such procedures are either restricted or forbidden. Second, the access to foetal tissue is limited, makin it difficult to obtain sufficient amounts for grafting. Third, with aborted foetal material there is a risk that an infectious agent can be transmitted with the graft tissue. Fourth, current protocols use transplant cells that express histocompatibility antigens. They are, therefore, immunogenic and may therfore require immunosuppression of the recipient.

Many of these problems could be overcome if alternative neurotransmitter producing cells could be maintained *in vitro* for intracerebral transplantation. In this context, we discuss experiments in which cells genetically modified to express tyrosine hydroxylase have been grafted into an animal model of PD.

Expression of tyrosine hydroxylase in mammalian cells

Tyrosine hydroxylase (TH), the key enzyme in dopamine synthesis, is a biopterin-dependent monoxygenase. The enzyme is homotetrameric and is localised in the cytoplasm of catecholaminergic cells. The gene encoding human TH generates four different RNAs through alternative splicing events (Grima et al. 1987; Le Bourdellès et al. 1988, 1991). At the protein level, the corresponding diversity is restricted to the N-terminal regulatory domain of the enzyme. In the following experiments, we used a cDNA encoding the most active isoform of the enzyme, designed HTH-I.

The HTH-1 human cDNA was introduced into various cell lines possessing clearly distinct features:

1. the NIH 3T3-fibroblastic cell line, which possesses a constitutive secretion pathway but no known regulated pathway for secretion;
2. the NS20 Y neuroblastoma cell line, which expresses choline acetyltransferase activity and exhibits many properties characteristic of differentiated neurons, in particular, induction of neurite outgrowth observed following serum withdrawal (Amano et al. 1972);
3. the AtT-20 cell line, which is derived from an anterior pituitary tumor, produces and stores corticotropin in secretory granules and releases it upon stimulation with secretagogues (Gumbiner and Kelly, 1981, 1982). This cell

line has provided an excellent model for studies on hormone secretion. Neurite-like outgrowths have also been observed in the presence of cAMP or laminin. We have also shown that this cell line contains the aromatic L-amino acid decarboxylase (AADC; Matsuuchi et al. 1988);
4. the RIN cell line, which is derived from pancreatic islet ß cells (Gazdar et al. 1980) and possesses the regulated secretory pathway and also expresses AADC.

These cells have been modified by retrovirus-mediated gene transfer or, in early experiments, by transfection. Following selection with neomycin, the resulting levels of TH activity in these cells were several hundred-fold lower than in the rat pheochromocytoma PC12 cell line (Greene and Tischler, 1982), in which TH represents about 0.1% of total proteins (Greene and Tischler, 1982). However, high levels of exogenous TH activity could be obtained after selection in tyrosine free medium (TFM). The increase in TH activity was nearly 1000 fold in the case of NIH 3T3. In the absence of tyrosine, this essential amino acid, which is required to produce L-DOPA, originates from the phenylalanine hydroxylase (PAH) activity associated with that of TH. The two hydroxylases, TH and PAH, which both utilize tetrahydrobiopterin as a cofactor, may substitute for each other, although with lower efficacy (Kaufman and Fischer, 1974). Therefore, the selection of cells according to their ability to synthesize tyrosine ensures that they efficiently produce tetrahydrobiopterin and L-DOPA. The selection in TFM is associated with an amplication of TH minigene (P. Horellou, unpublished data), a process most probably responsible for the increase in TH expression. Most importantly, the selection in TFM allowed a direct isolation of cells displaying *in vivo* TH activity.

The fate of L-DOPA differs among cell lines. The fibroblast and the neuroblastoma cells synthesized only LDOPA, which was abundantly secreted in the culture medium, presumably through a constitutive secretion pathway. In contrast, the neuroendocrine cell lines AtT-20 and RIN synthesized not only L-DOPA but also dopamine (DA), because of the presence of AADC in these cell lines. Comparison of the relative extracellular and intracellular concentrations of DA suggests that DA is stored in these cells (Horellou et al. 1989, 1990a). Moreover exposure to an external stimulus such as 56 mM potassium induces, as in PC12 cells, a calcium-dependent increase in DA release (Fig. 1). Hence, the additional expression of a single gene endows these endocrine cells with the capacity to synthesize and release a classical neurotransmitter in a regulated fashion. Further studies are needed to define the mechanisms of DA storage and release in both cell types.

Effects of L-DOPA- and DA-releasing cells on motor behaviour after implantation in the denervated striatum

To study the functional capacity of L-DOPA- and DA-producing cells in the brain, we analysed their effect on motor impairments after intrastriatal implantation in a rat model of PD (Ungerstedt 1968; Ungerstedt and Arbuthnott

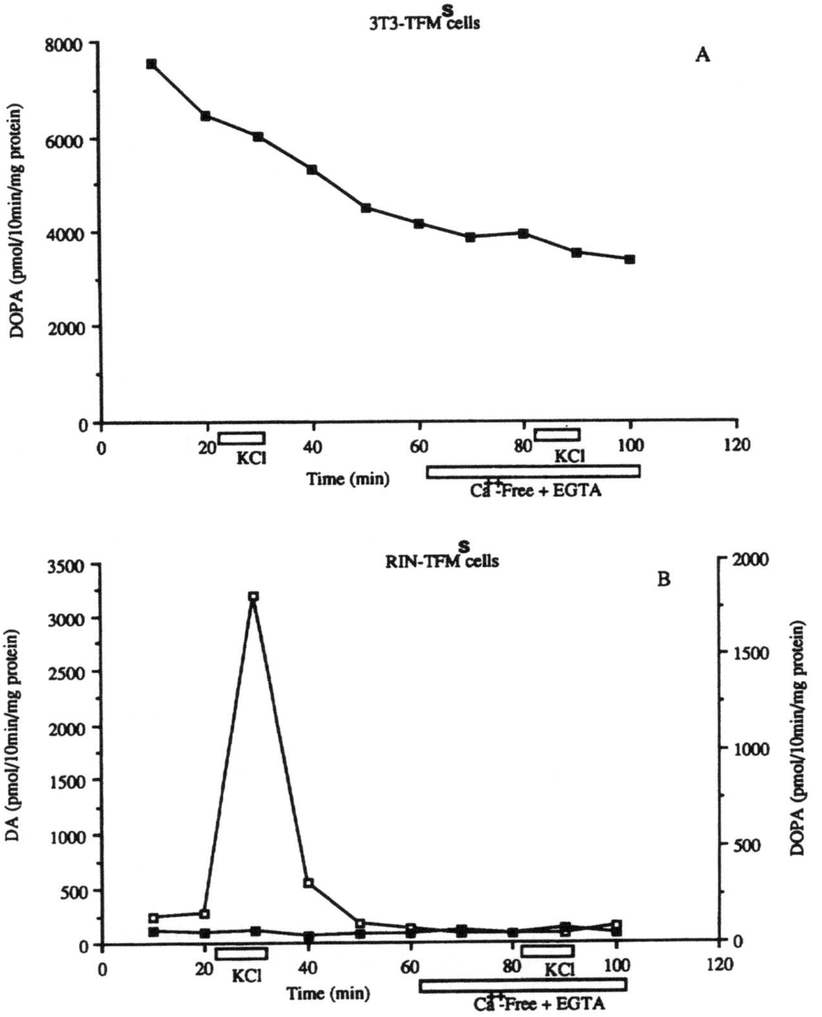

Fig. 1A, B. Release of DOPA by 3T3-TFM^{s} and RIN-TFM^{ss} cells in vitro, spontanously and in the presence of 56 mM potassium. The cultures were preincubated for 30 minutes in KRH medium containing 100 μM L-tyrosine. They were then washed and incubated with KRH medium, and the supernatant was removed every 10 minutes for analysis. KRH medium containing 56 μM potassium was applied after 20 minutes for 10 minutes. The effect of 56 μM potassium was also studied in the absence of calcium by using calcium-free KRH medium containing 1 mM EGTA. DOPA and DA levels were measured by HPLC (Horellou et al. 1989). The release of catechol is expressed in pmol/10 min per mg of protein. Each point represents the mean of two determinants. *A*, 3T3-TFM^{s}, *B*, RIN-TFM^{s}; note that DOPA values (closed squares) are given on a scale different from that of the DA values (open squares).

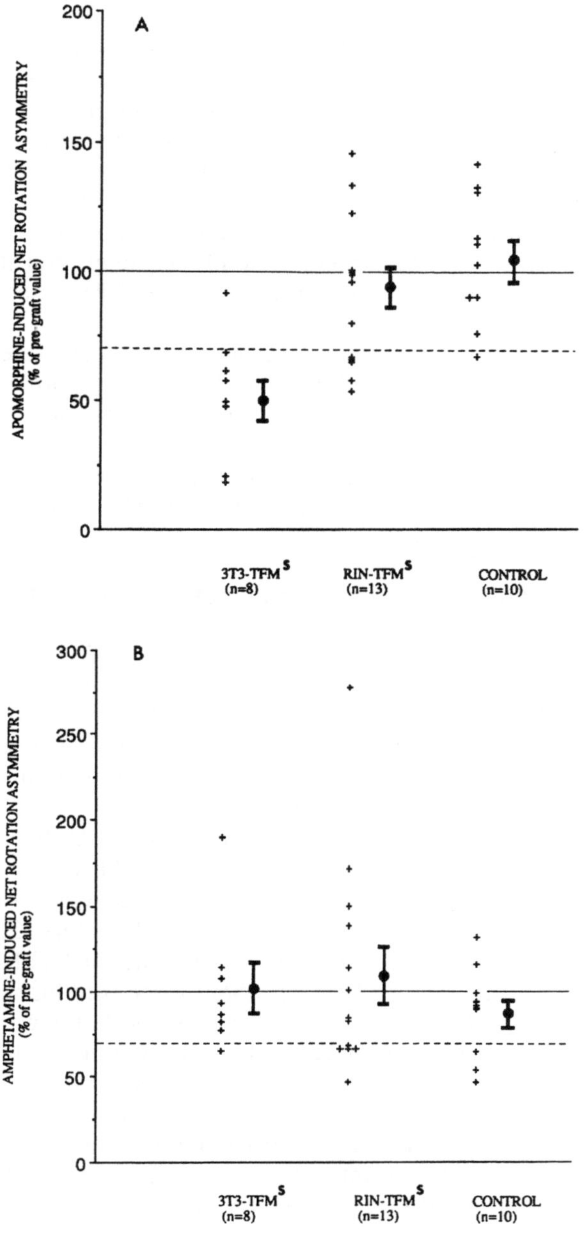

Fig. 2A, B. Turning behaviour. *A*, apomorphine-induced net rotation asymmetry (turns ipsilateral to lesion subtracted from turn contralateral to lesion, expressed as a percentage of the pretransplantation score) five days after grafting of 3T3-TFMs cells (n = 8), RIN-TFM^{s} cells (n = 13) and control cells (unmodified NIH 3T3 or RIN cells; n = 10). *B*, amphetamine-induced net rotation asymmetry (turns contralateral to lesionsubtracted fromturns ipsialteral to lesion, expressed as a percentage of the pretransplantation score) six days after grafting in the same groups of rats. Each symbol represents one animal. The mean and SEM values are also indicated (dots and bars). Solid lines and dotted lies indicate 100% and 70% values, respectively.

1970). In a first series of experiments we grafted the L-DOPA-producing neuroblastoma and the DA-producing pituitary cell lines. Unmodified cell lines wee used as controls. The rats, immunosuppressed with cyclosporin A (Brundin et al. 1985), where tested for apomorphine-induced turning at various time-points shortly after grafting. Intrastriatal grafts of AtT-20-TFM^s (producing DA) as well as of NS20 Y-TFM^s (producing DOPA) partially reversed the apomorphine-induced behaviour up to 50%. This effect was specific to the genetically modified cells, since it was not observed with grafts of unmodified cells. Immunohistochemical study of the grafted striatum showed that both cell lines survived and continued to express TH for at least two weeks (Horellou et al. 1990b).

A second series of experiments was performed with the fibroblastic 3T3-TFM^s and RIN-TFM^s cells producing DA. Both cell lines reversed the apomorphine-induced rotation but the effect was more significant with the L-DOPA producing DA (Fig. 2). In this instance the study was carried out further by analysing the capacity of these two cell lines to release catechols *in vivo*, as well as to convert L-DOPA to DA in the DA denervated striatum.

Microdialysis of catechols released from the grafts *in vivo*

The capacity of the 3T3-TFM^s and RIN-TFM^s cells to release catechols *in vivo*, to convert DOPA to DA and to release in the DA denervated striatum was analysed by microdialysis experiments performed eight to nine day after grafting. Dialysis probes were implanted in the host striatum at a distance of about 0.6 mm from each of the three implantation sites. Dialysis was started 12–18 hours after probe implantation.

The highest extracellular levels of L-DOPA (45 ± 8 fmol/25 μl) were observed around the 3T3-TFM^s grafts (Table 1; Fig. 3A), in agreement with the properties of these modified cells *in vitro*. In contrast to the situation *in vitro*, however, high extracellular levels of DA (20-fold higher than those of L-DOPA) were also present around the 3T3-TFM^s grafts (Fig. 3A). Furthermore, this level of DA was about three-fold higher than that found around the RIN-TFMS grafts, and the level of the latter was similar to that measured in normal rat striatum (Fig. 3C). Since not decarboxylation of L-DOPA was found *in vitro*, these findings suggest that the rleased DOPA was rapidly decarboxylated to DA by the host brain environment. This activity is most probably localized in nondopaminergic neuronal elements and in the endothelial vascular cells (Hefti et al. 1980; Melamed et al. 1980).

The level of DA in 3T3-TFM^s grafts resulting from an efficient conversion of DOPA might explain why these cells were more efficient that the RIN-TFM^s cells to reduce apomophine-induced turning, as assessed five to six days after grafting. Furthermore, L-DOPA- may be more efficient than DA in reaching larger areas of the denervated host striatum, since DA is more rapidly metabolized by monoamine oxidase than DOPA, which is a poor substrate for this enzyme. This would allow the diffusion of secreted DOPA over longer distances and thus provide a supply of DA over larger areas of the host striatum.

Table 1. Tissue content and basal extracellular levels of DOPA and DA measured in grafted and normal striata[1]

	3T3-TFMs (n = 5)	RIN-TFMs (n = 5)	N3T3 (n = 6)	Normal (n = 4)
DOPA				
P (fmol/25 μl)	45 ± 8	17 ± 6	19 ± 8	22 ± 9
T (Fmol/mg)	666 ± 280	213 ± 66	ND	9 ± 5
P/T (ratio)	0.068	0.8	–	2.56
DA				
P (fmol/25 μl)	842 ± 68	265 ± 57	17 ± 5	172 ± 13
T (fmol/mg)	390 ± 37	8710 ± 556	195 ± 23	15180 ± 844
P/T (ratio)	2.16	0.03	0.09	0.01

[1] Basal extracellular levels of catechols were measured in the dialysis perfusates and are expressed in fmol/25 μl. Tissue DOPA and DA levels measured postmortem in normal and grafted striata were assayed by the radioenzymatic method as described in the text and are expressed in fmol per mg of tissue. Each value is the mean ± SEM of the first 30-minute baseline samples taken from four to six rats, prior to addition of KCI (see Fig. 3). Perfusate-to-tissue (P/T) ratio corresponds to the relative ratio of extracellular (fmol/25 μl) to tissue (fmol/mg (DOPA) or DA levels. P, perfusate; T tissue.

Grafting of DOPA-secreting cells may therefore be functionally more efficient *in vivo* than that of DA-secreting ones, since the diffusion of DOPA through the striatum may provide a substrate for more efficient local DA synthesis, similar to that obtained after systemic L-DOPA administration.

The effect of depolarisation was also studied *in vivo* by adding a high concentration of potassium (100 mM), in the presence or absence of calcium. With 3T3-TFM^s grafts, high potassium did not result in peak evoked DA release (but only a two-fold increase with no decrease after removal of stimulation). This results is in agreement with the high perfusate-to-tissue concentration of DA in these cells (2.16 compared with 0.001 in the normal striatum), suggesting that newly synthesized DA is stored only to a very limited extent in the denervated striatum. The unexpected effect of calcium-free medium containing EGTA which increases L-DOPA- overflow in the grafted striatum is likely to be due to changes in DOPA handling or metabolism within the host striatum, since this effect was not seen when the 3T3-TFM^s cells were exposed to the same conditions *in vitro*.

In contrast, the striatum grafted with RIN-TFM^s cells and the normal intact striatum both displayed sharp increases in DA release following exposure to high potassium (13- and 70-fold increase, respectively; Fig. 3B and 3C). Perfusion with calcium-free medium (containing EGTA) reduced, but did not abolish, DA overflow in the normal and RIN-TFM^s grafted striata. We assume that the failure to block DA overflow completely is due to the difficulty in removing all free calcium from the area around the probe (Kalén et al. 1988). The result, together with the low perfusate-to-tissue ratio, confirms earlier *in*

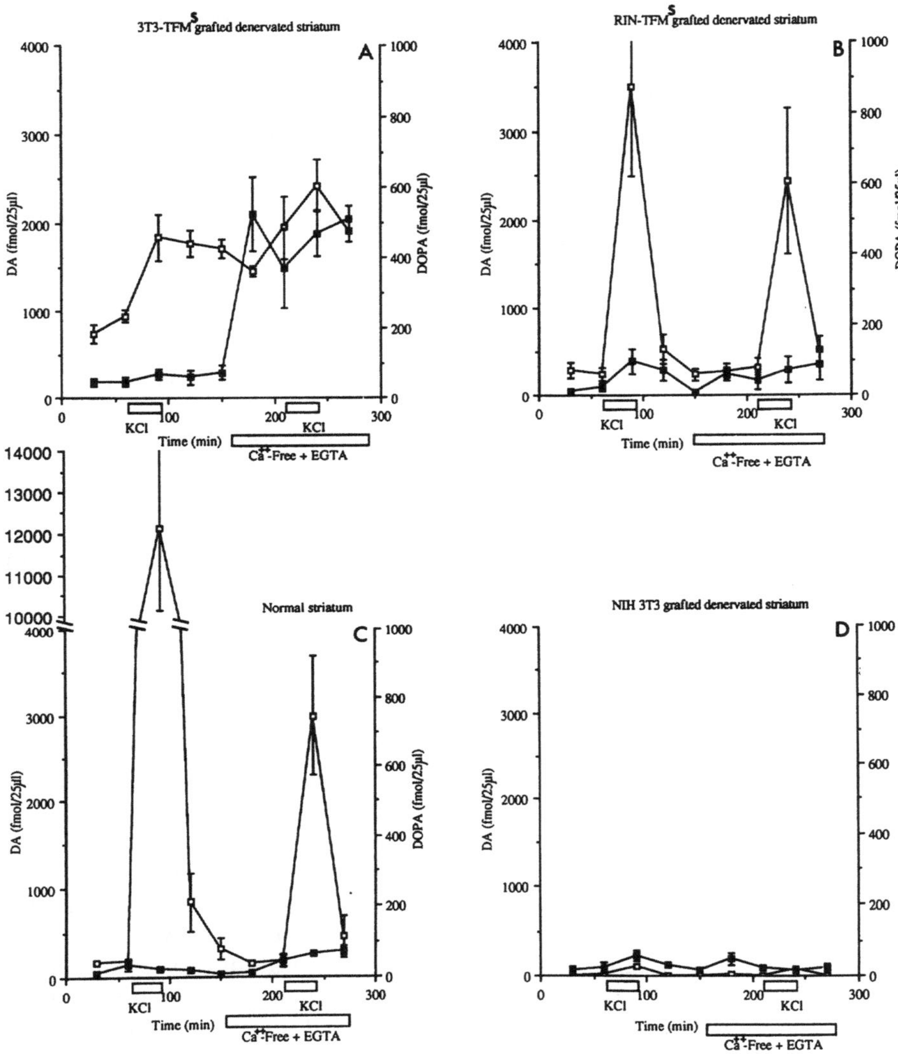

Fig. 3A–D. Levels of DOPA and DA in the dialysis perfusates from 6-OHDA-lesioned striata grafted with 3T3-TFM^s cells, RIN-TFMs cells, or unmodified NIH 3T3 control cells, and from normal intact striata. *A*, 3T3-TFM^s grafted denervated striatum (n = 5); *B*, RIN-TFM^s grafted denervated striatum (n = 5); *C*, normal intact striatum (n = 4); *D*, NIH 3T3 grafted denervated striatum (n = 6). Each 30-minute sample provided 60 µl of perfusate 2 µl/min perfusion rate. DOPA and DA were assayed by the radioenzymatic method described in the text. Levels are expressed in fmol/25 µl and are represented as mean ± SEM (bars). Note than the DOPA values (closed squares) are given on a four-fold larger scale than the DA values (open squares).

vitro experiment and provides evidence that the newly synthesised DA was stored in RIN-TFM^s cells and released in a regulated fashion.

None of the grafted cell types modified the amphetamine-induced turning response. This is consistent with our *in vitro* data (unpublished data) showing that catechol release from 3T3-TFM^s cells was not influency by amphetamine. In a previous study Wolff et al. (1989) reported that grafted rat fibroblasts genetically modified to produce L-DOPA in the presence of biopterin could reduce not only apomorphine- but also amphetamine-induced turning behaviour in grafted animals. However, the mechanism of this effect was not explored in the study.

Conclusion and perspectives

The experiments described above served to establish that genetically modified cells producing neurotransmitters (or L-DOPA) may be useful for compensating functional deficits following intraceerebral grafting. Over the short post-grafting period used here (10 days), the DOPA-secreting cells appeared to be functionally more efficient than the DA-secreting ones, probably because the DOPA-producing cells led to the highest extracellular levels of DA.This finding is consistent with previous DA infusion studies showing that locally applied DA in the striatum is functional only at high concentrations (Pijnenburg and Van an Rossum 1973; Jackson et al. 1975; Costall et al. 1977; Freed 1985; Hargraves and Freed 1987; Winn et al. 1989), reflecting the high levels of DA normally present in the synapses (Gonon and Buda, 1985).

Future experiments will focus on the use of primary cells or transformed cells that have kept the property of contact inhibition. These cells may be selected because they display characteristics such as coexpression of peptides (Hökfelt et al. 1980; Artaud et al. 1989), growth factors (Rosenberg et al. 1988; Ernfors et al. 1988), or the ability to extend neurites and establish synapses (Brundin et al. 1988; Clarke et al. 1988). Such cells should allow further analysis of mechanisms of action and transplant-induced functional recovery in experimental parkinsonism. One attractive possibility is the use of glial cells and, more particularly, astrocytes. These cells can be selected because they offer several advantages. They have been shown to survive after intracerebral grafting, and restricted subtypes of astroglial cells may migrate some distance from the implantation site (Lindsay and Raisman 1984). They may induce the formation of the blood-brain barrier (Janzer and Raff 1987); the also produce biopterin, the required TH cofactor. In the future, it may be possible to convert astrocytes to neurones making use of appropriate differentiation factors. Such a differentiation switch was recently described with the muscle cell differentiation program using the MyoD1 gene.

The use of recombinant retroviruses as vehicles to transfer genes encoding neurotransmitter synthesising enzymes into somatic cells is of particular interest because retrovirus allow efficient integration of a gene as a single, transcriptionally active copy in one chromosome site of dividing cells. A safe system sould be possible to develop to eliminate the risk of tumorigenesis and viral dissemination. While tumorigenesis may occur through insertional activation of cellular oncogenes after the integration of a retrovirus (Neel et al. 1981; Payne et al. 1981; Moreau-Gachelin et al. 1988) or retroposition (Heidmann et al. 1988), oncogene activation is a rare event which may be circumvented by the use of a defective LTR prometer (Yu et al. 1986; Cone et al. 1987; Dougherty and Tenin 1987). Viral dissemination can be avoided by using safe helper-free packaging cell lines (Danos and Mulligan 1988; Markowitz et al. 1988; Morgenstern and Land 1990).

Another issue concerns the stability of expression of host genes, since the activity of "exogenous promoters" such as those of viruses has been found to fade with time and may disappear after a few weeks. In this context, various endogenous active promoters, such as those driving the expression of enkephalin and GFAP, may be more appropriate.

With further developments along these lines, we are optimistic that the genetic engineering approach will be able to provide new sources of homogeneous, standardised, efficient ad safe neurotransmitter-producing cells for transplantation purposes.

References

Amano T, Richelson E, Niremberg M (1972) Neurotransmitter synthesis by neuroblastoma clones. Proc Natl Acad Sci USA 69:258–263

Artaud F, Baruch P, Stutzman JM, Saffroy M, Godeheu G, Barbeito L, Hervé D, Studier JM, Glowinski J, Chéramy A (1989) Cholecystokinin: co-release with dopamine from nigrostriatal neurons in the cat. Eur J Neurosci 1:162–171

Björklund A, Stenevi U (1979) Reconstruction of the nigrostriatal dopamine pathway by intracerebral nigral transplants. Brain Res 177:555–560

Björklund A, Dunett SB, Stenevi U, Lewis ME, Iversen SD (1980) Reinnervation of the denervated striatum by substantia nigra transplants: functional consequences as revealed by pharmacological and sensorimotor testing. Brain Res 199:307–333

Brundin P, Nilsson OG, Gage FH, Björklund A (1985) Cyclosporin A increases survival of cross-species intrastriatal grafts of embryonic dopamine-containing neurons. Exp Brain Res 60:204–208

Brundin P, Strecker RE, Lindvall O, Isacson O, Nilsson OG Barbin G, Prochiantz A, Forni C, Nieoullou A, Widner H, Gage FH, Björklund A (1987) Intracerebral grafting of dopamine neurons. Ann NY Acad Sci 495:473–496

Brundin P, Strecker RE, Widner H, Clarke DJ, Nilsson OG, Asted B, Lindvall O, Björklund A (1988) Human fetal dopamine neurons grafted in a rat model of Parkinson's disease: immunological aspects, spontaneous and drug-induced behaviour and dopamine release. Exp Brain Res 70:192–208

Clarke DJ, Brundin P, Strecker RE, Nilsson OG, Björklund A, Lindvall O (1988) Human fetal dopamine neurons grafted in a rat model of Parkinson's disease: ultrastructural evidence for synapse formation using tyrosine hydroxylase immuno cytochemistry. Exp Brain Res 73:115–126

Cone RD, Weber-Benarous A, Baorto D, Mulligan RC (1987) Regulated expression of a complete human β-globin gene encoded by a transmissible retrovirus vector Mol Cell Biol 7:887–897

Costall B, Naylor RJ, Cannon JG, Lee T (1977) Differentiation of the dopamine mechanisms mediating stereotyped behaviour and hyperactivity in the nucleus accumbens and caudate-putamen. J Pharm Pharmacol 29:337–342

Danos O, Mulligan RC (1988) Safe and efficient generation of recombinant retroviruses with amphotropic and ecotropic host ranges. Proc Natl Acad Sci USA 85:6460–6464

Dougherty JP, Temin HM (1987) A promoterless retroviral vector indicates that there are sequences in U3 required for 3′ RNA processing. Proc Natl Acad Sci USA 84:1197–1201

Dunnett SB, Björklund A, Stenevi U, Iversen SD (1981) Behavioural recovery following transplantation of substantia nigra in rats subjected to 6-OHDA lesions of the nigrostriatal pathway I. Unilateral lesions. Brain Res 215:147–161

Dunnett SB, Hernandez TD, Summerfield A, Jonses GH, Arbuthnott G (1988) Graft-derived recovery from 6-OHDA lesions: specificty of ventral mesencephalic graft tissues Exp Brain Res 71:411–424

Ehringer H, Hornykieewicz O (1960) Verteilung von noradrenalin and dopamin (3-hydroxytryramin) im Gehirn des Menschen und ihr Verhalten bei Erkrankungen des extrapyramidalen. Klin Wochenschr 38:1236–1239

Ernfors P, Ebendal T, Olson L, Mouton P, Strömberg I, Persson H (1988) A cell line producing recombinant nerve growth factor evokes growth response in intrinsic and grafted central cholinergic neurons. Proc Natl Acad Sci USA 86:4756–4760

Freed CR, Breeze RE, Rosenberg N et al. (1990) Transplantation of human fetal dopamine cell for Parkinsons's disease: results at one year. Arch Neurol 47:505–512

Freed WJ (1985) Transplantation of tissues to the cerebral ventricles: methological details and rate of graft survival. In: Björklund A, Stenevi U (eds) Neural grafting in the mammalian CNS. Amsterdam, Elsevier, pp 31–41

Gazdar AF, Chick WL, Oie HK, Sim HL, King DL, Weir GC, Lauris V (1980) Continuous clonal insulin- and somatostatin-secreting cell lines established from a transplantable rat islet cell tumor. Proc Natl Acad Sci USA 77:3519–3523

Gonon FG and Buda MJ (1985) Regulation of dopamine release by impulse flow and by autoreceptors as studied by in vivo voltammetry in the rat striatum. Neuroscience 14:765–774

Greene LA, Tischer A (1982) PC12 cultures in neurobiologial research. Adv Cell Neurobiol 3:373–414

Grima B, Lamouroux A, Boni C, Julien JF, Javoy-Agid F, Mallet J (1987) A single human gene encoding multiple tyrosine hydroxylase with different predicted functional characteristics. Nature 326:707–711

Gumbiner B, Kelly RB (1981) Secretory granules of an anterior pituitary cell line, AtT-20, contain only mature forms of corticotropin and β-lipotropin. Proc Natl Acad Sci USA 78:318–322

Gumbiner B, Kelly RB (1982) Two distinct intercellular pathways transport secretory and membrane glycoproteins to the surface of pituitary tumor cells. Cell 28:51–59

Hargraves R, Freed WJ (1987) Chronic intrastriatal dopamine infusions in rats with unilateral lesions of the substantia nigra. Life Sci 40:959–966

Hefti F, Melamed E, Wurtman RJ (1980) The decarobxylation of DOPA in the Parkinsonian brain: in vivo studies on an animal model. J Neural Transm 16:95–101

Heidmann T. Heidmann O, Nicolas JF (1988) An indicator gene to demonstrate intracellular transposition of defective retroviruses. Proc Natl Acad Sci USA 89:2219–2223

Herman JP, Choulli K, Le Moal M (1985) Hyper-reactivity to amphetamine in rats with dopaminergic grafts. Exp Brain Res 60:521–526

Herman JP, Lupp A, Abrous N, Le Moal M, Herting G, Jackisch R (1988) Intrastriatal dopaminergic grafts restore inhibitory control over striatal cholinergic neurons. Exp Brain Res 73:236–248

Hökfelt T, Skirboll L, Rehfeld JF, Goldstein M, Marey K, Dann O (1980) A subpopulation of mesencephalic dopamine neurons projecting to limbid areas contains a cholecystokinin-like peptide: evidence from immunohistochemistry combined with retrograde tracing. Neuroscience 5:2093–2124

Horellou P, Guibert B, Leviel V, Mallet J (1989) Retroviral transfer of a human tyrosine hydroxylase cDNA in various cell lines: regulated release of dopamine in mouse anterior pituitary AtT-20 cells. Proc Natl Acad Sci USA 86:7233-7237

Horellou P, Brundin P, Kalén P, Mallet J, Björklund A (1990a) In vivo release of DOPA and dopamine from genetically engineered cells grafted to the denervated rat striatum. Neuron 5:393–402

Horellou P, Marlier L Privat A, Darchen F, Scherman D, Henry JP, Mallet J (1990b) Exogenous expression of L-DOPA and dopamine in various cell lines following retroviral transfer of a human tyrosine hydroxylase cDNA: grafting in an animal model of Parkinson's disease. In: Dunnett S, Richard SJ (eds) Progress in brain research. 82:23–32

Jackson DM, Andén NE, Dahlström A (1975) A functional effect of dopamine in the nucleus accumbens and in some other dopamine rich parts of the rat brain. Psychopharmacologia 45:139–149

Janzer RC, Raff MCC (1987) Astrocytes induce blood brain barrier properties in endothelial cells. Nature 325:263–257

Kalén P, Strecker RE, Rosengren E, Björklund A (1988) Endogenous release of neuronal serotonin and 5-HIAA in the caudate-putamen of the rat as revealed by intracerebral dialysis coupled to high performance liquid chromatography with fluorimetric detection. J Neurochem 51:1422–1435

Kaufman S, Fischer DB (1974) Pterin-requiring aromatic amino acid hydroyxlase. In: Hayaishi O (ed) Molecular mechanism of oxygen activation. Academic Press, pp 285–369

Le Bourdellès B, Bourlarand S, Boni C, Horellou P, Dumas S, Grima B, Mallet J (1988) Analysis of the human tyrosine hydroxylase gene: combinatorial patterns of exon splicing generate multiple regulated tyrosine hydroxylase soforms. J Neurochem 50:988–991

Le Borudellès B, Horellou P, Le Caer JP, Denèfle P Latta M, Haavik J, Guibert B, Mayaux JF, Mallet J (1991) Phosphorylation of human recombinant tyrosine hydroxylase isoform 1 and 2: an additional phosphorylated residue in isoform 2, generated through alternative splicing. J Bio Chem 26:17124–17130

Lindvall O, Brundin P, Widner H, Rehncrona S, Gustavii B, Frackowiak R, Leenders KL, Sawle G, Rothwell JC, Marsden CD, Björklund A (1990) Grafts of fetal dopamine neurons survice and improve motor function in Parkinson's disease. Science 247:574–577

Lindsay RM, Raisman G (1984) An autoradiography study of neuronal development, vasçularization and glial cell migration from hippocampal transplants labelled in intermediate explant culture. Neuroscience 12:513–530

Mandel RJ, Brundin P, Björklund A (1990) The importance of graft placement and task complexity for transplant-induced recovery of simple and complex sensorimotor deficits in dopamine denervated rats. Eur J Neurosci 2:888–894

Markowitz D, Goff S, Bank A (1988) A safe packaging line for gene transfer: separating viral genes on two different plasmids. J Virol 62:1120–1124

Matsuuchi L, Buckley KM, Lowe AW, Kelly RB (1988) Targeting of secretory vesicles to cytoplasmic domains in AtT-20 and PC12 cells. J Cell Biol 106:239–251

Melamed E, Hefti F, Wurtman RJ (1980) Non-aminergic striatal neurons convert exogenous L-DOPA to dopamine in Parkinsonism. Ann Neurol 8:558–563

Moreau-Gachelin F, Tavitian A, Tambourin P (1988) Spi-1 is a putative oncogene in virally induced murine erythroleukaemias. Nature 331:277–280

Morgenstern JP, Land H (1990) Advanced mammalian gene transfer: high titre retroviral vectors with multiple drug selection markers and a complementary helper-free packaging cell line. Nucl Acids Res 19:3587–3596

Neel BG, Hayward WS, Robinson JL, Fang J, Astrin SM (1981) Avian leukosis virus-induced tumors have common proviral integration sites and synthesize discrete new RNAs: oncogenesis by promoter insertion. Cell 23:323–334

Payne SG, Courtneidge SA, Crittenden LB, Fadly AM, Bishop JM and Varmus HE (1981) Analysis of avian leukosis virus DNA and RNA in bursal tumors: viral gene expression is not required for maintenance of tumor state. Cell 23:311–322

Perlow MJ, Freed WJ, Hoffer BJ, Seiger A, Olson L, Wyatt RJ (1979) Brain grafts reduce motor abnormalities produced by destruction of nigrostriatal dopamine system. Science 204:643–646

Pijnenburg AJJ, Van Rossum JM (1973) Stimulation of locomotor activity following injection of dopamine into the nucleus accumbens. J Pharm Pharmacol 25:1003–1005

Rosenberg MB, Friedmann T, Robertson RC, Tuszynski M, Wolff JA, Breakefield XO, Gage FH (1988) Grafting genetically modified cells to damaged brain: restorative effects of NGF expression. Science 242:1575–1578

Schmidt RH, Ingvar M, Lindvall O, Stenevi U, Björklund A (1982) Functional activity of substantia nigra grafts reinnervating the striatum: neurotransmitter metabolism and [^{14}C] 2-deoxy-D-glucose autoradiography. J Neurochem 3:737–748

Schmidt RH, Björklund A, Stenevi U, Dunnett SB, Gage FH (1983) Intracerebral grafting of neuronal cell suspensions. III. Activity of intrastriatal nigral suspension implants as assessed by measurement of dopamine synthesis and metabolism. Acta Physiol Scand 522 (suppl): 19–28

Strecker RE, Sharp T, Brundin P, Zetterström T, Ungerstedt U, Björklund A (1987) Autoregulation of dopamine release and metabolism by intrastriatal nigral grafts as revealed by intracerebral dialysis. Neuroscience 22:169–178

Ungerstedt U (1968) 6-hydroxydopamine induced degeneration of central monoamine neurons Eur J Pharmacol 5:107–110

Ungerstedt U, Arbuthnott GW (1970) Quantitative recording of rotational behavior in rats after 6-hydroxydopamine lesions of the nigrostriatal dopamine system Brain Res 24:485–493

Winn SR, Wahlberg L, Tresco PA, Aebischer P (1989) An encapsulated dopamine-releasing polymeer alleviates experimental parkinsonism in rats. Exp Neurol 105:244–250

Wolff JA, Fisher LJ, Xu L, Jinnah HA, Langlais PJ, Luvone PM, O'Malley KL, Rosenberg MB, Shimohama S, Friedmann T, Gage FH (1989) Grafting fibroblasts genetically modified to produce L-DOPA in a rat model of Parkinson disease. Proc Natl Acad Sci USA 86:9011–9014

Yu SF, von Ründen T, Kantoff PW, Garber C, Seiberg M, Rüther U, Anderson FW, Wagner EF, Gilboa E (1986) Self-inactivating retroviral vectors designed for transfer of whole genes into mammalian cells. Proc Natl Acad Sci USA 83:3194–3198

Zetterström T, Brundin P, Gage FH, Sharp T, Isacson O, Dunnett SB, Ungerstedt U, Björklund A (1986) in vivo measurement of spontaneous release and metabolism of dopamine from intrasriatal nigral grafts using intracerebral dialysis. Brain Res 362:344–349

Neurotrophins and their Receptors

H. Persson, P. Ernfors, C. F. Ibañez, F. Hallböök,
W. J. Friedman, J.-P. Merlio, O. Lindvall, J. Bengzon,
N. Lindefors, T. Ebendal, and *L. Olson*

Summary

Development and maintenance of the mammalian nervous system require neurotrophic factors. The most well-characterizied neurotrophic factor, nerve growth factor (NGF), has recently been shown to be a member of a family of structurally related proteins that also includes brain-derived neurotrophic factor (BDNF), neurotrophin-3 (NT-3) and neurotrophin-4 (NT-4). An evolutionary study of these factors (neurotrophins) in representative groups of vertebrates showed that the four factors are highly conserved from fishes to mammals. Chimeric molecules were contructed where different combinations of sequences from BDNF replaced the corresponding sequences in NGF. Assays for biological activity of such molecules showed that the biological specificities of the two proteins are obtained by specific combinations of a set of sequences that differ between the two molecules. In the brain, the highest levels of NGF, BDNF and NT-3 mRNAs are found within the hippocampus, where they are expressed in a unique set of neurons. Both BDNF and NT-3 mRNAs are transiently expressed in some brain regions during postnatal brain development, suggesting that trophic interactions during brain development are a dynamic phenomenon which may alter with various stages of ontogeny. Expression of NGF and BDNF mRNAs in the brain appears to be regulated by neuronal activity, and the levels of these mRNAs show a marked and transient increase following epileptic seizures, cerebral ischemia and hypoglycemic coma. The increases of NGF and BDNF mRNAs appear to be mediated via a release of glutamate activating AMPA receptors within the hippocampus, as well as a rise in intracellular calcium levels. The proto-oncogene trkB, an essential component of the high-affinity BDNF receptor, is also expressed in the hippocampus, where it is spatially and temporally regulated in a manner similar to BDNF. These data suggest that BDNF and its receptor could, in a paracrine/autocrine fashion, play a role in kindling-associated neural plasticity and in neuronal protection following ischemic and hypoglycemic insults. Genetically modified cells that express high levels of the four neurotrophins have been established and used for implantation into the rat brain to identify cells that respond to these factors *in vivo*. Using this approach, NGF has been shown to expert a trophic stimulation for cholinergic neurons in the brain.

Introduction

In the developing vertebrate nervous system, neurons are produced in vast excess and, within a restricted time period following the arrival of axons in the target areas, a significant portion of them undergo naturally occurring cell death (Hamburger and Oppenheim 1982). The neurons that reach maturity are thought to have successfully competed for a target-derived, retrogradely transported neurotrophic factor present in limited amounts in the target fields (Barde 1989). In the peripheral nervous system, the most well-characterized neurotrophic factor, nerve growth factor (NGF), supports the development of sympathetic and neural crest-derived sensory neurons (Levi-Montalcini 1987). Confirming a trophic role of NGF for adult sympathetic neurons, the levels of both NGF mRNA and protein correlate with the density of sympathetic innervation (Heumann et al. 1984; Shelton and Reichardt 1984). More recently, NGF has also been shown to be expressed in the brain, where the highest levels of its mRNA have been found in the hippocampus and cerebral cortex (Korsching et al. 1985; Whittemore et al. 1986; Shelton and Reichardt 1986), areas receiving major cholinergic projections from the basal forebrain. Subsequently, *in situ* hybridization has been used to demonstrate that, within the hippocampus, NGF mRNA is expressed in neurons, implying that NGF exerts a trophic effect in the brain through neuron-to neuron interactions (Ayer-LeLievre et al. 1988).

NGF mediates its effects by interacting with specific receptors present on the surface of responsive neurons (Meakin and Shooter 1991; and references therein). The NGF-receptor (NGF-R) can occur in low and high affinity states (Landreth and Shooter 1980; Schechter and Bothwell 1981; Sutter et al. 1979), though only the high-affinity receptor mediates the biological activity of NGF. The low-affinity NGF-R is a 75 K glycosylated polypeptide and is expressed in NGF-responsive basal forebrain cholinergic neurons (Hefti et al. 1986; Richardson et al. 1986; Taniuchi et al. 1986). Chemical cross-linking experiments have demonstrated a NGF-R complex in NGF-responsive PC12 cells, consisting of the 75K low-affinity receptor and a second polypeptide of 140K (Meakin and Shooter 1991). The 140K component has recently been suggested to be the trk proto-oncogene (here designated trkA), (Kaplan et al.1991; Klein et al. 1991a). Coexpression of the trkA gene and the gene for the low-affinity NGF-R results in high-affinity NGF binding, possibly as a result of heterodimer formation between the trkA tyrosine kinase receptor and the 75K low-affinity NGF-R (Hempstead et al. 1991). However, activation of trkA by NGF has been shown to induce resting NIH 3T3 cells to enter S phase and become morphologically transformed, suggesting that trkA alone could mediated the effects of NGF (Cordon-Carda et al. 1991).

In addition to NGF, one other protein, brain-derived neurotrophic factor (BNDF), has been shown to support survival of neurons *in vivo* (Hofer and Barde 1988). In common with NGF, BDNF supports the survival of neural crest-derived embryonic sensory neurons *in vitro*, but non-overlapping trophic activites are suggested by the finding that BDNF also supports placode-derived neurons from the nodose ganglia and retinal ganglion cells (Lindsay et al.

1985), which are less sensitive to NGF (Davies et al.1986). NGF is synthesized as a preproprotein and the structure of both the precursor and the mature protein has been deduced from cDNA and genomic clones (Scott et al. 1983; Ullrich et al. 1983). More recently, a genomic clone has been isolated for porcine BDNF (Leibrock et al. 1989). Of considerable interest is the finding that the mature BNDF and NGF proteins show striking amino acid similarities, suggesting that they are structurally related and may be members of a family of neurotrophic factors.

In this article we summarize our results on the isolation of two novel neurotrophic factors structurally related to NGF, as well as the localization of cells expressing NGF and its related factors in the central nervous system (CNS) and the regulation of their synthesis in the brain. We also provide evidence for how these factors acquire their biological specificity and demonstrate that genetically modified cells expressing NGF and its related proteins can be used to identify cells in the brain that respond to these factors *in vivo*.

Results and Discussion

Molecular cloning and structure of two novel neurotrophins

A set of degenerate oligonucleotides, representing all possible codons corresponding to the amino acid sequence KQYFYET and WRFIRID shared between NGF and BDNF, were used to prime the amplification of rat hippocampal cDNA by the polymerase chain reaction (PCR). The amplified DNA included a product with significant similarity to NGF and BDNF, which was used to isolate a 1020 nucleotide-long cDNA from a rat hippocampal library. Nucleotide sequence analysis of this clone showed an open reading frame encoding a 282 amino acid-long protein (Ernfors et al. 1990a). The C-terminal part of this protein contained a potential cleavage site for a 119 amino acid protein with 57% amino acid similarity to both rat NGF and pig BDNF. Due to its restricted expression in the brain, which is mostly confined to a subset of pyramidal and granular neurons in the hippocampus (see below), we named this protein hippocampus-derived neurotrophic factor (HDNF). Independent of our work, five other groups have also reported the molecular cloning of the gene for this protein and have named it neurotrophin-3(NT-3; Hohn et al. 1990; Maisonpierre et al. 1990; Rosenthal et al. 1990) or NGF-2 (Kaisko et al. 1990).

To obtain information that would allow a reconstruction of the evolutionary history of the distinct NGF-like genes, the same pool of degenerate oligonucleotides ware used to clone the genes for NGF, BDNF and NT-3 from seven different species representing different groups of vertebrates (Hallböök et al. 1991). The data obtained from this study allowed us to construct phylogenetic trees for NGF, BDNF and NT-3, where the relationship of each member of the family to the other members was studied. The phylogram showed that NGF is more closely related to NT-3 than to BDNF and that NT-3 is as related to NGF as to BDNF. A comparison of NGF sequences from different species with the

```
NT-4 ASGSDSVSLSRRGELSVCDSVNVWVT--DKRTAVDDRGKIVTVMSEIQTLTGPLKQYFFET
NGF      .STHPVFHM..F.......SV..G--..T..T.IK..E...LA.VNINNSVFR......
BDNF       HSDPAR.........ISE...AA..K....MS.GT...LEKVPVSK.Q.....Y..
NT-3      YAEHK.H...Y.......ESL...--..SS.I.I..HQ...LG..K.GNS.V....Y..

KCNPSGSTTRGCRGVDKKQWISECKAKQSYVRALTIDANKLVGWRWIRIDTACVCTLLSRTGRT 123
..RASNPVES.....I.S.H.N.Y.TTTHTF.K...T.E-.QAA..F.........V.SRKAT.  118
.....M.Y.KE.....I..RH.N.Q.RTT.......M.SK.RI...F.....S.......TIKR..  119
R.KEARPVKN.....I.D.H.N.Q..TS.T.....SEN.............S....A.SRKI...  119
```

Fig. 1. Amino acid (single letter code) sequence comparison of Xenopus NT-4 with mouse NGF (Scott et al.1983), mouse BDNF (Hofer et al. 1990) and mouse NT-3 (Hohn et al. 1990). Identical amino acid replacements compared with the NT-4 amino acid sequence are shown by dots. Sequences that differ between NGF, BDNF and NT-3 also differ in the sequence of the NT-4 protein. The six conserved cysteine residues involved in the formation of disulphide bridges are shown in boxes

corresponding sequences of BDNF or NT-3 showed that the NGF gene is less conserved in vertebrates than either BDNF and NT-3. The two latter genes appear to be equally conserved in all species studied, except in salmon, in which NT-3 is less conserved than BDNF.

Our study of the evolutionary conservation of the NGF family also led to the isolation of a novel member of this family, named neurotrophin-4 (NT-4; Hallböök et al. 1991). PCR fragments from the NT-4 gene were isolated from Xenopus and viper. Nucleotide sequence analysis of the NT-4 clone from Xenopus revealed an open reading frame for a 236 amino acid protein, which showed several structural features resembling those of the three other members of the NGF family. These including the presence of a putative amino-terminal signal sequence and a potential N-glycosylation site close to a porteolytic cleavage site that predicts a 123 amino acid mature NT-4 protein.

An alignment of the mature protein sequence of the four factors to each other showed that all six cysteine residues involved in the formation of disulphide bridges are conserved (Fig. 1). Each of the four proteins differs from the other members of the family in the same regions. Seven regions can be identified where the amino acid sequences of three or more residues are identical between the four proteins. An overall amino acid identify of 51–60% is seen between the four mature neurotrophins.

Biological acitivities of the neurotrophins and structural elements determining their specificites

Plasmid constructs containing the four neurotrophin genes, respectively, were used to transfect COS cells. Conditioned media were then tested for biological activity in bioassays that measured fiber outgrowth from various chicken embryo ganglia. In this assay, NT-3 was shown to cause a marked stimulation of neurite outgrowth in the ganglion of Remark (Ernfors et al. 1990a), a ganglionated nerve trunk in the mesotectum of the chicken embryo (Le

Douarin et al. 1978). NGF was far less efficient than NT-3 in stimulating fiber outgrowth from the Remark ganglia. In the paravertebral sympathetic trunk ganglia, NGF caused a massive fiber outgrowth, whereas only a weak response was seen with NT-3. Both BDNF and NT-3 stimulated fiber outgrowth in the placode-derived nodose ganglion, whereas no acitivty was seen with NGF. Similar results were reported by Maisonpierre et al (1990), also using the explanted chicken embryo ganglia assays. Similary to BDNF, Xenopus NT-4 stimulated fiber outgrowth from the nodose ganglia but showed no activity in the sympathetic ganglia (Hallböök et al. 1991). All four neurotrophins promoted fiber outgrowth in the sensory dorsal root ganglia. Cominbed, these results show that the four neurotrophins stimulate fiber outgrowth both from overlapping and unique sets of neuronal populations.

Structural elements determining the biological specificites of NGF and BDNF were examined using site-directed mutagenesis. Five variable regions in which the BNDF sequence differs significantly from that of NGF were defined (Fig. 2A). Chimeric molecules were then constructed where the variable regions in the NGF molecule were systematically replaced by the corresponding sequences from BDNF. The chimeric molecules were then transiently expressed in COS cells and equivalent amounts of the different recombinant proteins were tested for their ability to promote neurit outgrowth from embryonic chick sympathetic ganglia (SYMP; characteristic of NGF activity), nodose ganglia (NOD; characteristic of BDNF activity) or spinal dorsal root ganglia (SPIN; both NGF and BDNF activities). No significant change in the trophic activities was seen when the five regions were replaced independently in the NGF molecule. However, when combined, regions III and V resulted in a significant stimulation of neurite outgrowth from the NOD in addition to wild-type (wt) NGF activities in both SYMP and SPIN (Ibañez et al. 1991). The activity in NOD was further enchanced by adding region IV to the III + V molecule. The resulting chimera (III + IV + V) reached levels of activity in NOD comparable to those seen with wt BDNF while retaining at least half of the wt NGF activity in SYMP and SPIN (Fig. 2B). The chimera I + III + IV + V also displayed high levels of activity in both SYMP and SPIN as well ans NOD (Ibanez et al. 1991). The fact that I + III + IV + V molecule only retains variable region II from NGF suggests that this region is responsible for the SYMP activity seen with this molecule. This is further suggested by the fact than the addition of region II from BDNF to the I + III + IV + V molecule resulted in a loss of SYMP activity, indicating that region II is important for maximal stimulation of SYMP by NGF. The gain of activity in NOD caused by combining carboxy-terminal regions from BDNF into NGF indicates that the carboxy-terminal sequences of BDNF (notably regions III and V) determine the specificity of BDNF activity in NOD.

The fact that the activity in NOD increased gradually from I and III + V molecules to the I + III + IV + V molecule indicates that the interconversion of neurotrophic profiles in the NGF family is not an "all or nothing" phenomenon determined by a restricted group of amino acid residues. Instead, the specificity for a defined profile of activity appears to depend more on the overall shape or conformation of the molecule. As shown by the chimeric

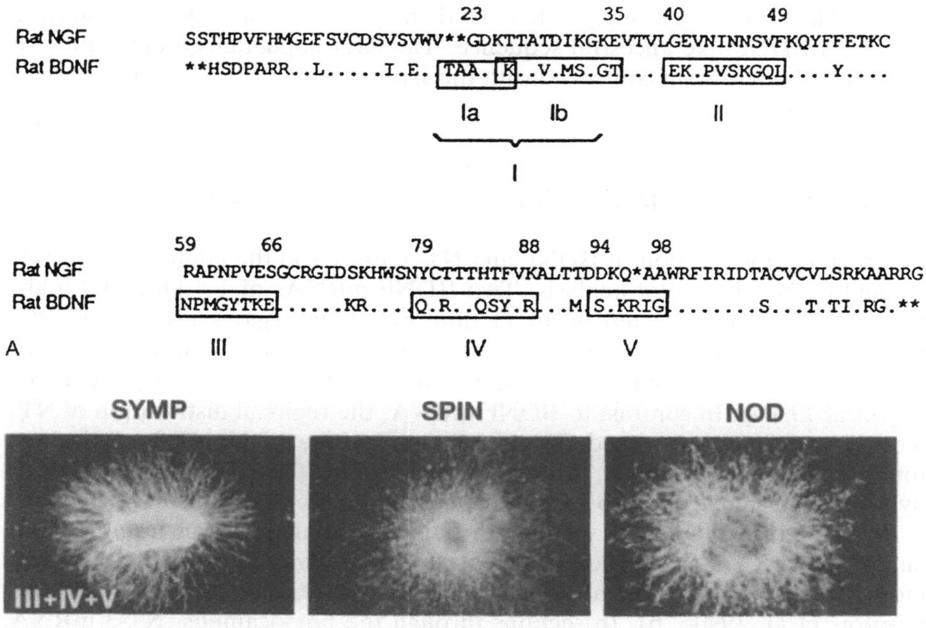

Fig. 2 A, B. *A,* Alignment of the amino acid sequences (single-letter code) of rat NGF (Whitemore et al. 1988) and rat BDNF (Maisonpierre et al. 1990). Boxes indicate variable regions in the BDNF sequence. In the different chimeric molecules, these regions were inserted in the same position in the NGF protein, thereby replacing the corresponding NGF sequences with the different variable regions from BDNF. Numbers indicate amino acid residues in the NGF sequence. *B,* Biological activies of chimeric molecule III+IV+V as assayed by neurite outgrowth stimulation of explanted chick embryonic day 9 ganglia. Dark-field photomicrographs of SYMP, SPIN and NOD cultured for 48 hours in the presence of COS cell conditioned media containing 2 ng/ml (SYMP and SPIN) or 8 ng/ml (NOD) of recombinant chimeric molecule III+IV+V

molecules I + IV + V and I + III + IV + V, some of these combinations allow a broader spectrum of neurotrophic activities than the two wild type proteins, clearly demonstrating that the activity profiles of NGF and BDNF are not mutually exclusive and can therefore be recruited in the same molecule. Chimeric proteins with a broad range of neurotrophic activites could represent molecules that combine the putative therapeutic values of both NGF and BDNF in neurodegenerative diseases.

The additive nature of the replacements allowed us (Ilbañez et al. 1991) to converge of the biological property in much the same way as proteins evolve, by cycles of natural variation and selection. Our data indicate that more changes are necesarry to gain a BDNF activity profile in a NGF sequence than vice versa, suggesting a direction in which the evolution of these factors may have occured. This observation, together with the fact that BDNF sequences from different species are remarkably more conserved than those of NGF

(Hallböök et al. 1991), suggest that NGF may be derived from a duplicated copy of a BDNF-like ancestral sequence. The later appearance in evolution of the sympathetic system compared with the sensory systems is in agreement with this speculation.

Expression of neurotrophin mRNAs in the central nervous system

The regional distribution of BDNF and NT-3 mRNA in the adult rat brain was studied by Northem blot analysis. Two BDNF mRNAs of 1.4 kb and 4.0 kb, respectively, were detected in all 11 different brain regions tested, with the highest levels in hippocampus, cerebral cortex and pons. The lowest levels of BDNF mRNA were found in striatum, septum and medulla oblongata (Ernsfors et al.1990a). In contrast to BDNF mRNA, the regional distribution of NT-3 mRNA was more restricted. The highest level of the 1.3 kb NT-3 mRNA was found in the hippocampus, with a lower level in the cerebellum. A weak hybridization signal was also seen in the cerebral cortex (Ernfors et al. 1990a).

In situ hybridizitation of sections through the adult rat brain using BDNF and NT-3 mRNA-specific oligonucleotide probes was performed to obtain more precise information about the regional distribution of these mRNAs (Ernfors et al. 1990a, b). In sections through the hippocampus, NT-3 mRNA expressing cells were found medially and laterally in the pyramidal layer, constituting the medial aspect of CA1 and all of CA2 (Fig. 3A). Cells in CA3 and the hilar region showed no labeling for NT-3 mRNA (Fig. 3A). More caudally, medially located CA1 cells did not label, although CA2 cells also expressed high levels of NT-3 mRNA at this level. The overall intensity of labeling appeared to decrease in more caudal sections of the hippocampus. BDNF mRNA-expressing cells were detected primarily in CA2, CA3, and the hilar cells of the dentate gyrus (Fig. 3B). Lower levels of expression were also seen in CA1. Sections hybridized to a NGF mRNA-specific probe revealed NGF mRNA-expressing cells in all areas of the hippocampus (Fig. 3C). Intensely labeled cells were seen in the hilar region of the dentate gyrus. Strongly labeled cells were found within the pyramidal layer, but also outside this layer, predominantly in the stratum oriens (Fig. 3C). In contrast, NT-3 as well as BDNF mRNA-labeled cells were seen primarily in the pyramidal layer (Fig. 3A and B). A comparison of the hippocampal labeling patterns for the three

Fig. 3A–C. Dark-field micrographs of NT-3, BDNF, and NGF mRNA-expressing cells in the hippocampus. Coronal sections were hybridized to probes specific for NT-3 (*A*), BDNF (*B*), or NGF (*C*) and exposed to photographic emulsion. Note the difference in distribution of labeled cells between the three factors: the NT-3-specific probe labeled CA1 and CA2; the BDNF-specific probe labeled CA2, CA3 and the hilar region with lower labeling in CA1; and the NGF-specific probe labeled scattered neurons in the hilar region, pyramidal layer and stratum oriens. Arrowheads in *C* indicate some of the labeled cells in the stratum oriens; closed arrows indicate labeled cells in the pyramidal layer of the hippocampus, and open arrows indicate labeled hilar cells. Abbreviations: IG, induseum griseum; dg, dentate gyrus; O, stratum oriens; P, pyramidal layer; R, stratum radiatum; Hi, hilar region; CA1m, medial part of region CA1. Bar in (*C*), 400 μm; bar applies also to (*A*) and (*B*)

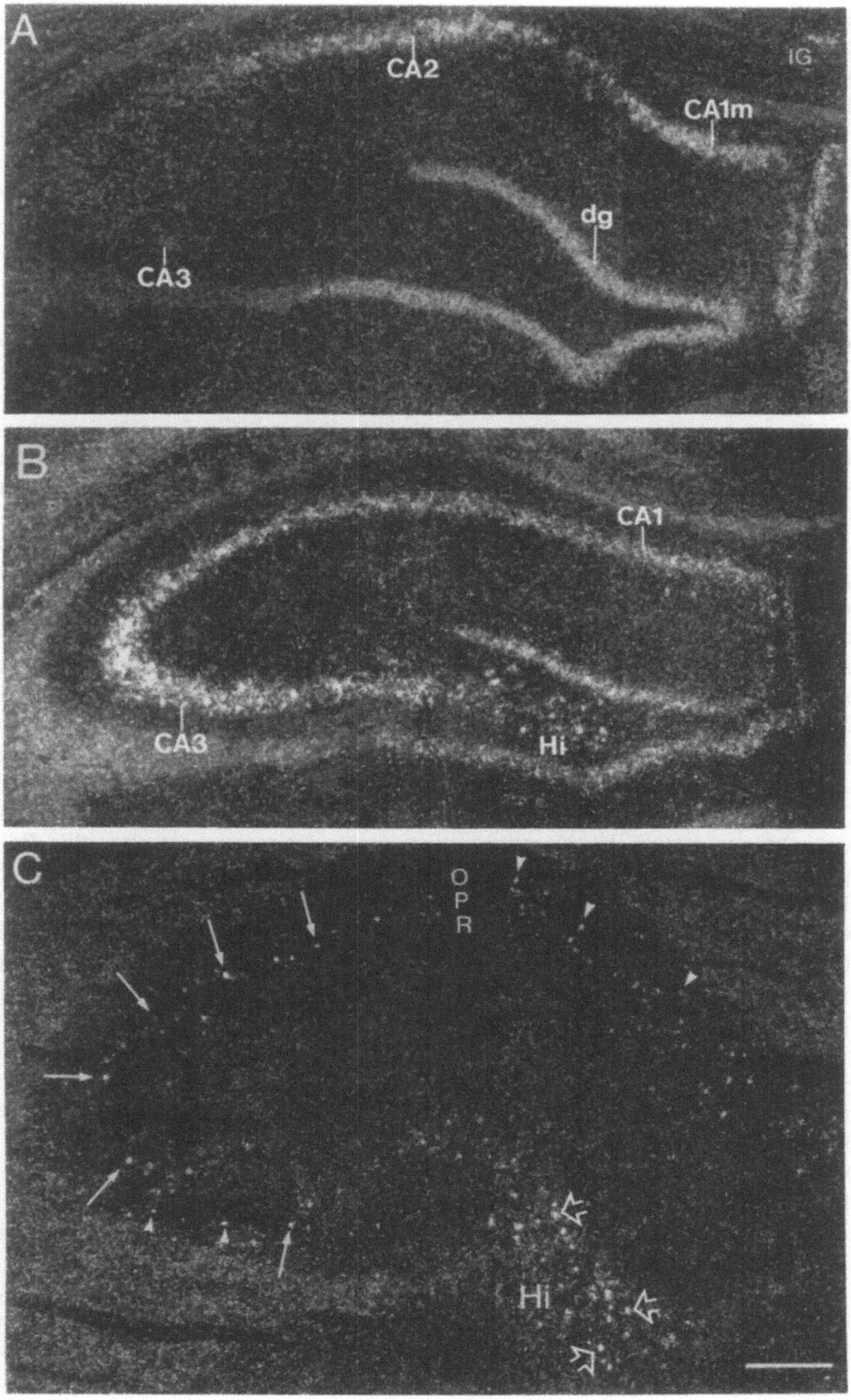
A
CA2
IG
CA1m
dg
CA3
B
CA1
CA3
Hi
C
O
P
R
Hi

different probes showed that NT-3, BDNF, and NGF are mostly expressed in areas of the hippocampus which are in many cases different from each other. NT-3 and BDNF mRNA expression was also seen over granular neurons of the dentate gyrus (Fig. 3), and a greater number of neurons was labeled for NT-3 mRNA than for BDNF mRNA (Figs. 3A and B). Few cells in the dentate gyrus labeled for NGF mRNA (Fig. 3C). Emulsion autoradiography revealed that the labeling in the hippocampus with any of the three different probes was over neurons. Some of the NGF mRNA-labeled neurons had a morphology of pyramidal neurons, and most NT-3 of BDNF mRNA-labeled neurons appeared to be pyramidal neurons (Ernsfors et al. 1990b).

Outside the hippocampal formation, NT-3 mRNA-labeled neurons were found only in a region of the brainstem, located medial and ventral to the locus coeruleus and the fourth ventricle, respectively and in the taenia tecta and induseum griseum (Ernfors et al. 1990b). In all these areas, the intensity of labeling was lower than that over intensely labeled NT-3 mRNA-expressing neurons in areas CA1 and CA2 of hippocampus. The induseum griseum and the taenia tecta are both developmentally part of the hippocampus (Swanson et al.1987), further emphasizing the strong assocation of NT-3 mRNA-expressing cells with the hippocampal formation.

In contrast to the restricted neuroanatomical arrangement to NT-3 mRNA-expressing cells in the brain, cells expressing BDNF mRNA were seen in several different brain regions (Ernsfors et al. 1990b). Intensely BDNF mRNA-labeled neurons were seen in the external and internal pyramidal layers of the cerebral cortex as well as in the claustrum. Less intensely labeled BDNF mRNAs-expressing neurons were also found in several other brain areas, including the anterior olfactory nucleus, taenia tecta, piriform and entorhinal cortices, and mammillary complex, as well as over cells in one brainstem area. The widespread distribution of BDNF mRNA-labeled cells is in agreement with the results of the Northern blot analysis and suggest that the BDNF protein is expressed in many different regions of the adult rat brain. Thus BDNF may support neuronal innervation in several different brain areas. The distribution of BDNF mRNA expressing cells, as revaled by *in situ* hybridization, has also been reported by us in the pig brain Wetmore et al. (1990) and by Hofer et al. (1990) and Philips et al. (1990) in the mouse and rat brain, respectively.

BDNF mRNA-expressing cells were also detected in the adult rat dorsal root ganglia (Ernfors et al. 1990b). The labeling was confined to the cytoplasm of neurons of various sizes. The intensity of labeling varied significantly among different neurons, and some neurons in the ganglia were not labeled. In the developing rat spinal cord the BDNF mRNA-specific probe labeled neurons in the E13, E15 and E19 dorsal root ganglia (Ernsfors and Persson 1991). The labeling was confined to the dorsal root ganglia with no labeling in the lateral motor column or in the rest of the spinal cord, including the white matter. The BDNF mRNA labeling was present over cells with a size and morphology of neurons although only a subpopulation of these was labeled. BDNF has been shown to stimulate fiber outgrowth and support survival of embryonic dorsal root ganglia neurons *in vitro* (Barde et al. 1982; Lindsay et al. 1985, Davies et al. 1986) and administration of BDNF to the quail embryo promotes survival of dorsal root ganglia neurions *in vivo* (Hofer and Barde 1988). The finding that a subset of dorsal root ganglia neurons expresses BDNF mRNA throughout development, at least from embryonic day 13 to adulthood, suggests a neurotrophic interaction which is different from the target-derived model. Thus, in both the developing and adult dorsal root ganglia, BDNF may stimulate dorsal root ganglion neurons by a local mechanicsm where the factor is synthesized in the same neuron as the one that requires the factor. Trophic stimulation can then either occur by release of BDNF, followed by interaction with its receptor on the cell surface, or by an intracellular association between BDNF and its receptor in a manner similar to that proposed for an autocrine stimulation of cell growth by platelet-derived growth factor (Heldin and Westermark 1990). Alternatively, BDNF locally produced in the dorsal root ganglia may function in a paracrine fashion, in which case the factor is released from one neuron and stimulates survival by interacting with its receptor present on neurons in close proximity within the dorsal root ganglia.

Northern blot analysis was used to demonstrate that NGF, BDNF and NT-3 mRNAs are maximally expressed in the brain at different times of development, with a peak of NT-3 mRNA shortly after birth, followed by maximal expression of BDNF and NGF mRNAs two and three weeks after birth, respectively (Ernsfors et al. 1990a). Thus, the three proteins may predominantly support neuronal innervation at different times of brain development and exert specific trophic support for different central nervous system neurons.

Using *in situ* hybridization the peak of NT-3 mRNA was shown to be due to a high and transient expression in cells in the cingulate cortex during the first two postnatal weeks, after which it is absent from that region, suggesting the existence of afferents which are transiently dependent upon NT-3 (Friedman et al. 1991a). BDNF mRNA was also detected transiently in several brainstem structures, notably in the substantia nigra and interpeduncular nucleus (Friedman et al. 1991b). The transient expression of BDNF mRNA in substantia nigra is particularly intriguing, as a trophic role for BDNF has been indicated for the dopaminergic neurons in this brain region (Hyman et al. 1991; Knüsel et al. 1991). This suggests the possibility that BDNF, similar to what has been proposed in the dorsal root ganglia (Ernfors et al. 1990b; Ernfors and Persson 1991), may function in an autocrine or local manner during early development in this region.

Regulation of neurotrophin mRNA expression in the brain

In the adult rat brain, the levels of NGF and BDNF mRNAs increase in the hippocampus, amygdala and neocortex after chronic seizures produced by a unilateral electrolytic lesion of the dentate gyrus hilus (Gall and Isackson 1989; Isackson et al. 1991).

Intraventricular, intracerebral or systemic administration of kainic acid also causes a marked but transient increase of NGF and BDNF mRNAs in the hippocampus and neocortex (Ballarin et al.1991; Gall et al. 1991; Zafra et al.

1990). This increase appears *in vitro* to be due to the release of glutamate interacting with non-NMDA receptors, probably of the kainate subtype (Zafra et al.1990). Moreover, activation of afferents from the medial septum to the hippocampus by injection of quisqualate into the medial septum leads to a marked but transient increase of BDNF and NGF mRNA expression in a subset of hippocampal neurons, and this increase can be completely prevented by pretreating the animals with the muscarinic antagonist scopolamine (Lindefors et al.1991). From the data obtained after chonric seizures and after injection of kainic acid or quisqualate, it has been hypothesized that the synthesis of NGF and BDNF in the brain is regulated by neuronal activity and stimulated by glutamate release.

Kindling, induced by repeated, often electrical, subconvulsive stimuli, leads to long-lasting facilitation of synaptic transmission and has been extensively studied as a model for development of hyperexcitability underlying seizure generation and spread in epilepsy (Morell and De Toledo-Morell 1986). Kindling leads to permanent changes in synaptic organization and fiber architecture, as demonstrated by the collateral sprouting of mossy fibers induced by kindling stimulations in the hippocampus (Messenheimer et al. 1979; Sutula et al. 1988). To examine the possibility that the collateral sprouting seen after kindling could be due to altered levels of the neurotrophins, expression of these mRNAs was studied during kindling epileptogenesis.

Two hours after one electrical stimulation in the ventral hippocampus, computerized image analysis of autoradiograms obtained after *in situ* hybridization revealed a 4.2-fold increase in the level of BDNF mRNA and a 1.3-fold elevation in NGF mRNA, compared with rats either connected to the stimulating-recording device in the absence on an electrical stimulus or subjected to low-frequency stimulation (Ernfors et al. 1991).

For both mRNAs, the increase was confined to granule cells of the dentate gyrus. The same increase was seen in both the stimulated and the nonstimulated hippocampi. The levels of NGF and BDNF mRNAs in the granule cells of the dentate gyrus were 1.5- and 1.6-fold higher, respectively, after 40 electrical stimulations than after one stimulation. Thus, two hours after the last stimulation the expression of BDNF mRNA was markedly elevated bilaterally in the dentate gyrus and, in contrast to animals receiving only one stimulation, there was also a bilateral increase throughout regions CA1-CA3 and in the hilar region of the dentate gyrus (Fig. 4), the amygdaloid complex, the piriform cortex, and the outer pyramidal layer of the parietal cortex (Ernfors et al. 1991). An increased level of NGF mRNA was detected in the dentate gyrus at two hours, but no change was seen in the pyramidal layer (Fig. 4). Four hours after the last stimulation, NGF mRNA was markedly increased in the piriform cortex and in the outer pyramidal layer of the parietal cortex. At 24 hours after the last stimulation, BNDF and NGF mRNAs has essentially returned to baseline. The level of NT-3 mRNA did not change in any of the experimental paradigms (Fig. 4). The increased expression of NGF and BDNF mRNAs after one electrical stimulation was not influenced by pretreatment with the NMDA receptor antagonist MK801, but was partielly blocked by the quisqualate, AMPA receptor antagnoist NBQX (Ernfors et al. 1991).

We also performed a study designed to explore the possibility that the expression of mRNAs for NGF, BDNF and NT-3 can be altered in response to ischemic and hypoglycemic insults which, similar to epileptic seizures, lead to an increase in glutamate release and to a rise in the intracellular, cytosolic calcium concentration. Levels of mRNA for NGF, BDNF and NT-3 were studied by *in situ* hybridization following forebrain ischemia induced by bilateral common carotid artery occlusion and simultaneous arterial hypotension. Increased levels of BDNF mRNA could be observed bilaterally over dentate gyrus granule cells starting at 30 minutes after two minutes of ischemia (Lindvall et al. 1992). The highest expression of BDNF mRNA was seen Two hours after the insult, when quantitative image analysis of autoradiograms showed a 3.2-fold increase compared to sham-operated animals. The level then tapered off although at 24 hours after the insult it was still 80% higher than the levels measured in controls. Increased levels of mRNA for NGF in the dentate gyrus granule cells were detected two hours after two minutes ischemia, and the highest expression (2.2-fold increase) was seen at four hours. Twenty-four hours after the insult, NGF mRNA expression had returned to control level. No changes in mRNA for BDNF and NGF were detected outside the dentate gyrus following ischemia. NT-3 mRNA expression was reduced in dentate gyrus granule cells and in the CA 2 and medial CA 1 regions of the hippocampus at two and four hours after a two-minute period of ischemia. The maximal reduction in the dentate gyrus, 61%, was seen four hours after two minutes of ischemia. At 24 hours NT-3 mRNA had returned to control levels (Lindvall et al. 1992). The increase of BDNF and NGF mRNA could be partially blocked by the AMPA receptor antagonist NBQX but was not influenced by the NMDA receptor antagonist MK-801. Both NBQX and MK-801 attenuated the decrease of NT-3 mRNA after ischemia. One and 30 minutes of hypoglycemic coma also induced marked increases in BDNF and NGF mRNA in dentate granule cells, with maximal levels at two hours (Lindvall et al. 1992).

In view of the marked changes of NGF and BDNF mRNAs during kindling epileptogenesis (Ernfors et al. 1991a) and following ischemia and hypoglycemic coma (Lindvall et al. 1992), it was of interest to examine the expression of members of the trk gene family encoding essential components of the high-affinity neurotrophin receptors during these conditions. To allow an analysis of the expression of members of the trk family, PCR cloning was used to isolate parts of the rat trkA and trkB genes encoding essential components of the high-affinity NGF (Kaplan et al. 1991; Klein et al. 1991a; Hempstead et al.1991) and BDNF (Klein et al. 1991b; Squinto et al. 1991; Soppet et al. 1991) receptor, respectively. A 2172 base pair long cDNA clone encoding 672 amino acids of rat trkC, recently shown to be an essential component of the high-affinity NT-3 receptor (Lamballe et al. 1991), was also isolated (Merlio et al., submitted for publication).

Hybridization of sections through the hippocampus from sham-operated adult rats revealed trkB mRNA in the granule cell layer of the dentate gyrus, the hilar region, throughout the pyramidal cell layer, the piriform cortex, neocortex, amygdaloid complex, striatum and in the ependymal lining of the

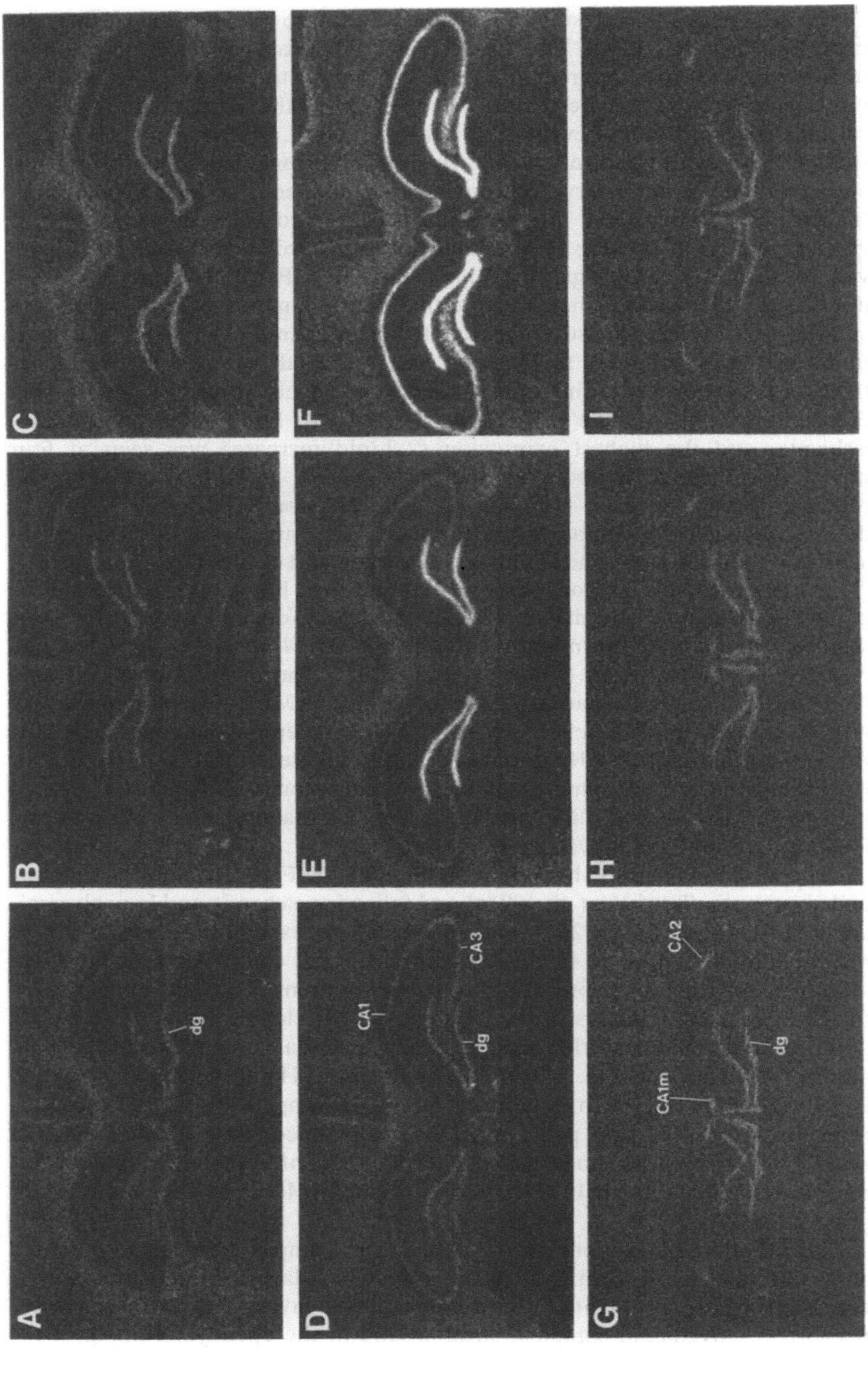

Fig. 4 A–I.

third and lateral ventricles and the lateral choroid plexus (Merlio et al., submitted for publication). The trkC mRNA-specific probe labeled cells in the granule cell layer of the dentate gyrus, in the hilar region,throughout the pyramidal layer of the hippocampus, the neocortex with highest intensity in layers II and III, the piriform cortex and the amygdaloid complex. The trkA mRNA-specific probe did not reveal any labeling over the hippocampus, though it detected trkA mRNA-expressing cells in the nucleus basalis of Meynert and in the medial septum (Merlio et al., submitted for publication). The distribution of trk A mRNA-exressing cells was very similar to the distribution of cells expressing mRNA for thelow-affinity NGF receptor (LNGFR), though the labeling was weaker with the trkA probe compared to the LNGFR probe.

Two hours after one electrical stimulation, the level of trkB mRNA showed a bilateral 2.5-fold increase in the granule cell layer of the dentate gyrus, whereas no change was seen in other regions (Merlio et al. submitted for publication). The level of trkA mRNA in the hippocampus also remained below the detection limit after the seizure, and no change was seen in the basal forebrain. In addition, the level of trkC mRNA in the hippocampus did not change after a single seizure (Merlio et al., submitted for publication). Thirty minutes after 40 electrical stimulations, the level of trkB mRNA was elevated 3.3-fold in the dentate gyrus. A less pronounced but clear increase was also observed throughout the pyramidal layer, the hilar cells of the hippocampus and in the piriform cortex. In all cases, the increase was seen on both the stimulated and contralateral side. The level of trkB mRNA remained elevated one and two hours after the last stimulation, but had declined at four hors. At 24 hours after the last stimulation, trkB mRNA had returned to the same level as seen in control animals. In contrast, no change was seen in the levels of trkA mRNAs in any of the brain areas studied at any of the time points tested (30 minutes and 1, 2, 4, 24 hours) after the last stimulation. A slight, although not statistically significant, increase in trkC mRNA was seen in the pyramidal layer of the hippocampus. The increase of trkB mRNA after one electrical stimulation was blocked by the AMPA receptor antagonist NBQX. TrkB mRNA was also transiently increased in the dentate gyrus following cerebral ischemia and hypoglycemic coma, whereas these treatments hat no effect on trkA and trkC mRNA (Merlio et al., submitted for publication).

Fig. 4A–I. Expression of mRNAs for neurotrophic factors in the dorsal hippocampus two hours after generalized of focal seizures. Coronal sections were prepared from control animals connected to the stimulating-recording device without electrical stimulation (*A, D, G*), from animals sacrificed after 40 electrical stimulatins (*C, F, I*) and from animals receiving only one kindling stimulation (*B, E, H*). The sections were hybridized to NGF (*A, B, C*) BDNF (*D, E, F*) or NT-3 (*G, H, I*) mRNA-specific probes. The figure shows prints of autoradiograms. Note the marked increase, restricted to the dendate gyrus, in NGF mRNA labeling after 40 stimulations (*C*) and the moderate increase seen after one stimulation (*B*). Also note the increased BDNF mRNA labeling in dentate gyrus and throughout the pyramidal layer (CA1-CA3 and in the hilus of the dentate gyrus) after 40 stimulations (*F*). After one electrical stimulation the increase was confined to the dentate gyrus (*E*). Abbreviations: dg, dentate gyrus: CA1, CA2 and CA3, regions CA1, CA2 and CA3 of the hippocampus.

The combined results of the studies of the expression of the neurotrophins and their receptors after seizures, ischemia and hypoglycemia coma imply that BDNF and trkB mRNA expression is regulated by similar mechanisms. Thus, a release of endogenous glutamate activating excitatory receptors of the AMPA subtype may cause an activity-dependent increase of both BDNF and trkB mRNA expression. Hypoglycemia and ischemia are, in contrast to seizures, assoicated with energy deprivation, which may lead to higher intracellular calcium levels, supposedly through ion influx via multipe pathways and release from intracellular stores (Siesjö and Bengtsson 1989) which may, at least in part, cause the increase in BDNF and trkB mRNA expression. Alternatively, the increase of trkB mRNA could be mediated via a ligand-induced activation, where an elevated level of the BDNF protein increases expression of mRNA encoding its own high-affinity receptor.

The distrubtion of trkB and trkC mRNA-expressing cells in the hippocampus and the regulation of their expression showed a striking similarity to the expression and regulation of BDNF and NT-3 mRNA. The high numbers of trkB and trkC as well as of BDNF and NT-3 mRNA-expressing cells in the granule layer of the dentate gyrus and the widespread distribution of such cells in the pyramidal layer of the hippocampus are consistent with the notion that some, or perhaps many, of these cells express both the neurotrophin and its receptor. If so, this implies an autocrine trophic stimulation. A paracrine trophic stimulation, were two neighboring cells express the neurotrophin and its receptor, respectively, is also a possible mechanism for mediating a trophic stimulation within the hippocampus. Such a neurotrophic support could be important fo kindling-associated plasticity in specific neuronal systems in the hippocampus, which could promote hyperexcitability and contribute to the development of epileptic syndromes. Our data also suggest that an autocrine/paracrine stimulation of neuronal survival by neurotrophins in the hippocampus could influence functional outcome and neuronal neurosis following ischemic and hypoglycemic insults.

The use of genetically modified fibroblasts to identify cells in the brain that respond to neurotrophins

To make possible a study of the effects of NGF on CNS neurons *in vivo*, genetically modiffied mouse 3T3 fibroblasts producing recombinant NGF were established. For this purpose, a DNA fragment from the 3'exon of the rat NGF gene encoding a prepro NGF protein was inserted into the mammalian expression vector MRE-(4x)-OVEC (Westin et al. 1988). In this construct, the rat NGF gene is transcribed from a rabbit β-globin promoter in conjunction with a metal ion-inducible enhancer from the mouse metallothionein I gene (Fig. 5A). This construct was cotransfected with a neomycin-resistance plasmid into mouse fibroblasts 3T3 cells. Southern blot analysis of genomic DNA from one transfected cell line (3E) revealed the presence of several hundred copies of the rat NGF gene. This cell line also expressed a 2.2 kB NGF-hybridizing transcript (Ernfors et al. 1989). Conditioned medium from the 3E cells readily

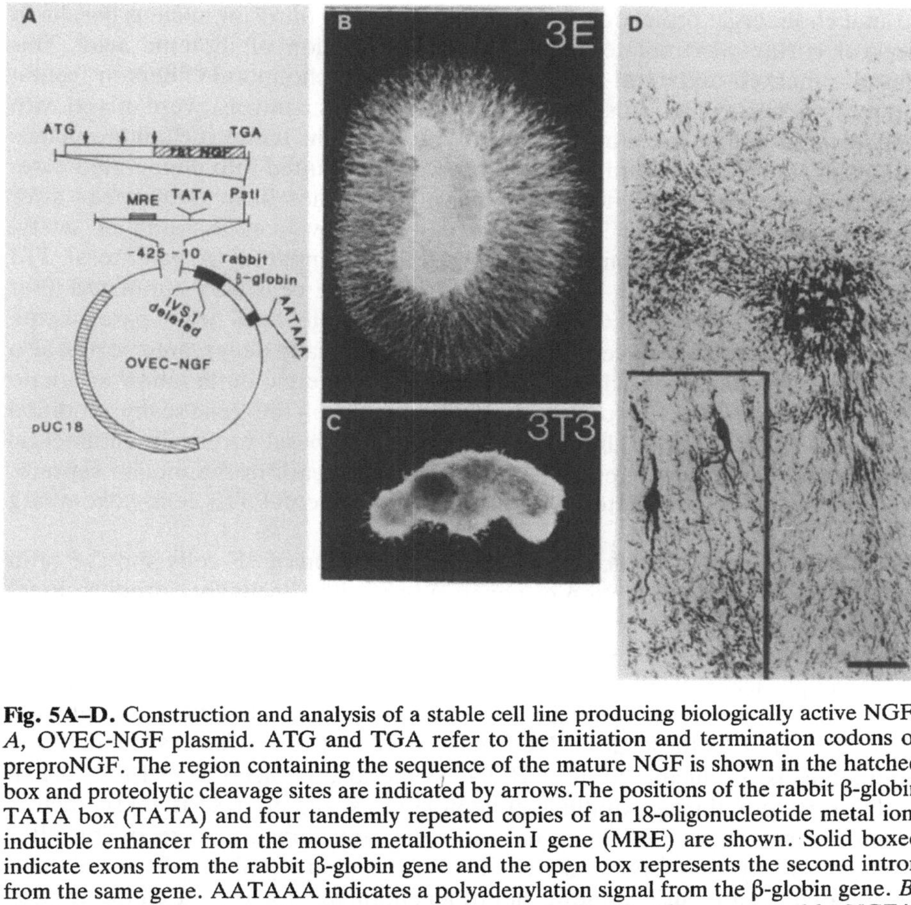

Fig. 5A–D. Construction and analysis of a stable cell line producing biologically active NGF. *A,* OVEC-NGF plasmid. ATG and TGA refer to the initiation and termination codons of preproNGF. The region containing the sequence of the mature NGF is shown in the hatched box and proteolytic cleavage sites are indicated by arrows.The positions of the rabbit β-globin TATA box (TATA) and four tandemly repeated copies of an 18-oligonucleotide metal ion-inducible enhancer from the mouse metallothionein I gene (MRE) are shown. Solid boxed indicate exons from the rabbit β-globin gene and the open box represents the second intron from the same gene. AATAAA indicates a polyadenylation signal from the β-globin gene. *B,* Conditioned medium from 3E cells or from the parental 3T3 cells *(C)* was assayed for NGF in a bioassay using chicken sympathetic ganglia. Note the dense nerve fiber halo formed with the medium from the 3E cell line. *D,* Growth responses evoked after implantation of cells producing recombinant NGF into the rat brain. AChE histochemistry of a coronal section from cholinergically denervated cerebral cortes four weeks after grafting of a dissociated cell suspension of embryonic day 17–18 basal forebrain cholinergic neurons mixed with 3E cells. Note marked innervation by AChE-positive fibers in graft area (dark densely stained area at midright) as well as around grafted neurons (e.g., midleft) and in adjacent cortical neuropil. Inset, two grafted AChE-positive neurons with several nerve fibers formed in response to 3E cells.

stimulated fibre outgrowth from chicken sympathetic ganglia (Fig. 5B). A two-site enzyme immunoassay for NGF (Lärkfors and Ebendal 1987) revealed 5± 1 ng of NGF protein per ml of conditioned medium from 3E cells compared to 0.2 ± 0.04 ng/ml produced by the parental cells.

Next, the NGF-producing 3E cells were used for implantation into the rat brain. As a model of the cholinergic deficits of Alzheimer's disease, the frontoparietal cholinergic projection from the anterior two-thirds of nucleus basalis to cerebral cortex was unilaterally lesioned by injection of ibotenic acid. This caused a marked unilateral reduction in density of cholinergic fibers in frontal and parietal cortex. 3E cells, or parental 3T3 cells as controls, were mixed with a dissociated cell suspension of fetal basal forebrain tissue including nucleus basalis magnocellularis, and the mixtures were implanted into denervated cerebral cortices. Counts of cholinergic nerve cell bodies from four weeks after implantation showed that 3E cells increased survival, as assessed by acetyl cholinesterase (AChE) staining, about 10-fold compared to parenteral 3T3 cells (Fig. 5D). In addition to increased survival, 3E cells also stimulated fiber outgrowth from grafted AChE-positive neurons into host brain parenchyma (Fig. 5D). Implants of 3E cells alone into cholinergically denervated cortex also significantly increased the density of intrinsic cholinergic innervation in a halo around the implants (Ernfors et al. 1989). Grafts of the genetically modified NGF-producing 3E cells also rescued axotomized basal forebrain cholinergic neurons and significantly reduced cholinergic cell death in the medial septum, as compared with rats treated with grafts of the parental 3T3 cells (Strömberg et al. 1990).

The marked neurotrophic effects after implantation of 3E cells into the ratin brain suggest that implantation of genetically modified cells that produce functional recombinant gene products (transgenes) into the rat brain can be used as a tool to unravel functional aspects of the product encoded by the transgene. Similar approaches using genetically modified cells producing BDNF, NT3- ot NT-4, now in progress, could be expected to identify cells in the brain that respond *in vivo* to these three NGF-related neurotrophic factors. Genetically modified cells producing NGF or BDNF could also be used to test the role of NGF or BDNF during kindling epileptogenesis and to test if these factors influence the degree of neuronal necrosis and functional recovery after ischemic and hypoglycemic insults. Finally, the neurotrophic role *in vivo* and the putative therapeutic value in neurodegenerative diseases of chimeric neurotrophin molecules with multiple neurotrophic activites *in vitro* (Ibañez et al. 1991) could be evaluated using this approach.

Acknowledgments. This work was supported by US grants (University of Colorado) AG 04418,NS 09199, and grants from The Swedish Natural Science Research Council, The Bank of Sweden Tercentenary Foundation, The Swedish Society Against Cancer, Swedish Research Council for Engineering Sciences, The Swedish Medical Research Council, Petrus and Augusta Hedlunds Stiftelse, Magnus Bergvalls Stiftelse, Thorsten and Elsa Segerfalks Stiftelse and funds from the Karolinska Institute. P.E was supported by a fellowship from the Swedish Medical Research Council. C.F.I. was supported by a fellowship from the Wenner-Gren Foundation. J.P.M. was supported by a French-Swedish fellowship from the Institut National de la Santé et de la Recherche Médicale-Swedish Medical Research Council.

References

Ayer-LeLievre C, Olson L, Ebendal T, Seiger A, Persson H (1988a). Expression of the α-nerve growth factor gene in hippocampus neurons. Science 240:1339–1341

Ballarin M, Ernfors P, Lindefors N, Persson H (1991) Hippocampal damage and kainic acid injection induce a rapid increase in mRNA for BDNF and NGF in the rat brain. 114:35–43

Barde Y-A (1989) Trophic factors and neuronal survival. Neuron 2:1525–1534

Barde Y-A, Edgar D, Thoenen H (1982) Purification of a new neurotrophic factor from mammalian brain. EMBO J 1:549–553

Cordon-Cardo C, Tapley P, Jing S, Nanduri V, O'Rourke E, Lamballe F, Kovary K, Klein R, Jones KR, Reichardt LF, Barbacid M (1991) The trk tyrosine protein kinase mediates the mitogenetic properties of nerve growth factor and neurotrophin-3. Cell 66:173–183

Davies AM, Thoenen H, and Barde YA (1986) The response of chick sensory neurons to brain-derived neurotrophic factor. J Neurosci 6:1987–1904

Ernfors P, Ebendal T, Olson L, Mouton P, Strömberg I, Persson H (1989) A cell line producing recombinant nerve growth factor evokes growth responses in intrinsic and grafted central cholinergic neurons. Proc Natl Acad Sci USA 86:4756–4760

Ernfors P, Ibanez CF, Ebendal T, Olson L, Persson H (1990a) Molecular cloning and neurotrophic activites of a protein with structural similarities to nerve growth factor: developmental and topographical expression in the brain. Proc Natl Acad Sci USA 87:5454–5458

Ernfors P, Wetmore C, Olson L, Persson H (1990b) Identification of cells in rat brain and periphereal tissues expressing mRNA for members of the nerve growth factor family. Neuron 5:511–526

Ernfors P, Bengzon J, Kokaia Z, Persson H, Lindvall O (1991) Increased levels of messenger RNAs for neurotrophic factors in the brain during kindling epileptogenesis. Neuron 7:165–176

Ernfors P, Persson H (1991) Developmentally regulated expression of HDNF/NT-3 mRNA in rat spinal cord motoneurons and expression of BDNF mRNA in embryonic dorsal root ganglion. Eur J Neurosci 3:953–961

Friedman WJ, Ernfors P, Persson H (1991a) In situ hybridization reveals both transient and persistent expression of HDNF/NT-3 mRNA in the rat brain during postnatal development. J Neurosci 11:1577–1584

Friedman WJ, Olson L, Persson H (1991b) Cells that express brain-derived neurotrophic factor mRNA in the developing postnatal rat brain. Eur J Neurosci 3:688–697

Gall CM, Isackson PJ (1989) Limbic seizures increase neuronal production of messenger RNA for nerve growth factor. Science 245:758–761

Gall C, Murray K J IP (1991) Kainic acid-induced seizures stimulate increased expression of nerve growth factor mRNA in rat hippocampus. Mol Brain Res 9:113–123

Hallböök F, Ibañez CF, Persson H (1991) Evolutionary studies of the nerve growth factor reveal a novel member abundantly expressed in Xenopus ovary. Neuron 6:845–858

Hamburger V, Oppenheim RW (1982) Naturally occurring neuronal death in vertebrates. Neurosci Commun 1:39–55

Hefti F, Hartikka J, Salvatierra A, Weiner WJ, Mash DC (1986) Localization of nerve growth factor receptors in cholinergic neurons of the human basal forebrain. Neurosci Lett 69:37–41

Hempstead BL, Martin-Zanca D, Kaplan DR, Parada LF, Chao MV (1991) High-affinity NGF binding requires coexpression of the trk proto-oncogene and the low-affinity NGF receptor. Nature 350:678–683

Heumann R, Korsching S, Scott J, Thoenen H (1984) Relationship between levels of nerve growth factor (NFGF) and its messenger RNA in sympathetic ganglia and peripheral target tisues. EMBO J 3:3183–3189

Hofer MM, Barde Y-A (1988) Brain-derived neurotrophic factor prevents neuronal death in vivo. Nature 331:261–262

Hofer M, Pagliusi SR, Hohn A, Leibrock J,Barde YA (1990) Regional distribution of brain-derived neurotrophic factor mRNA in the adult mouse brain. EMBO J 9:2459–2464

John A, Leibrock J, Bailey K, Barde YA (1990) Identification and characterization of a novel member of the nerve growth factor/brain-derived neurotrophic factor family. Nature 344:339–341

Hyman C, Hofer M, Barde Y, Juhasz M, Yancopoulos GD, Squinto SP, Lindsay RM (1991) BDNF is a neurotrophic factor for dopaminergic neurons of the substantia nigra. Nature 350:230–232

Kaplan DR, Martin-Zanca D, Parada LF (1991) Tyrosine phosphorylation and tyrosine kinase activity of the trk protooncogene product induced by NGF. Nature 350:158–160

Ibañez E, Ebendal T, Persson H (1991) Chimeric molecules with multiple neurotrophic activities reveal structural elements determining the biological specificities of NGF and BDNF . EMBO J 10:2105–2110

Isackson PJ, Huntsman MM, Murray KD, Gall CM (1991)BDNF mRNA expression in increased in adult forebrain after limbic seizures: Temporal patterns of induction distinct from NGF. Neuron 7:937–948

Kaisho Y, Yoshimura KKN (1990) Cloning and expression of a cDNA encoding a novel human neurotrophic factor. FEMS 266:187–191

Kaplan DR, Hempstead BL, Martin-Zanca D, Chao MV, Parada LF (1991) The trk proto-oncogene product: a signal transducing receptor for nerve growth factor. Science 252:554–558

Klein R, Jing SQ, Nanduri V, O'Rourke E, Barbacid M (1991a) The trk proto-oncogene encodes a receptor for nerve growth factor. Cell 65:189–197

Klein R, Nanduri V, Jing S, Lamballe F, Tapley P, Bryant S, Cordon-Cardo C, Jones KR, Reichardt LF, Barbacid M (1991b) The trkB tyrosine protein kinase is a receptor for brain-derived neurotrophic factor and neurotrophin-3. Cell 66:395–403

Knüsel B, Winslow JW, Rosenthal A, Burton LE, Seid DP, Nikolics K, Hefti F (1991) Promotion of central cholinergic and dopaminergic neuron differentiation by brain-derived neurotrophic factor but not neurotrophin 3. Proc Natl Acad Sci USA 88:961–965

Heldin CH, Westermark B (1990) Platelet-derived growth factor: mechanism of action and possible in vivo function. Cell Regulation 1:555–566

Korsching S, Auberger G, Heumann R, Scott J, Thoenen H (1985) Levels of nerve growth factor and its mRNA in the central nervous system of the rat correlate with cholinergic innervation. EMBO J 4:1389–1393

Lamballe F, Klein R, Barbacid M (1991) trkC, a new member of the trk family of tyrosine protein kinases, is a receptor for neurotrophin-3. Cell 66:967–979

Landreth GE, Shooter EM (1980) Nerve growth factor receptors on PC12 cells: ligand-induced conversion from low- to high-affinity states Proc Natl Acad Sci USA 77:4751–4755

Larkfors L, Ebendal T (1987) Highly sensitive enzyme immunoassays for β-nerve growth factor. J Immunol Meth 26:41–47

Le Douarin NM, Teillet MA, Ziller C, Smith J (1978) Adrenergic differentiation of cells of the cholinergic ciliary and Remark ganglia in avian embryo after in vivo transplantation. Proc Natl Acad Sci USA 75:2030–2034

Leibrock J, Lottspeich F, Hohn A, Hofer M, Hengerer B, Masiakowski P, Thoenen H, Barde Y-A (1989) Molecular cloning and expression of brain-derived neurotrophic factor. Nature 341:149–152

Levi-Montalcini R (1987) The nerve growth factor: Thirty-five years later. Biosci Rep 7:681–699

Lindefors N, Ernfors P, Falkenberg T, Persson H (1992) Septal cholinergic afferents regulate expression of brain-derived neurotrophic factor and β-nerve growth factor mRNAs in rat hippocampus. Exp Brain Res 88:78–90

Lindsay RM, Thoenen H, Barde Y-A (1985) Placode and neural crest-derived sensory neurons are responsive at early developmental stages to brain-derived neurotrophic factor. Dev Biol 112:319–328

Lindvall O, Ernfors P, Bengzon J, Kokaia Z, Smith M-L, Siesjö BK, Persson H (1992) Differential regulation of mRNAs for nerve growth factor, brain-derived neurotrophic factor and neurotrophin-3 in the adult rat brain following cerebral ischemia and hypoglycemic coma. Proc Natl Acad Sci USA 89: 648–652

Maisonpierre PC, Belluscio L, Squinto S, Ip NY, Furth ME, Lindsay RM, D YG (1990) Neurotrophin-3: A neurotrophic factor related to NGF and BDNF. Science 247:1446–1451

Meakin SO, Shooter EM (1991) Molecular investigations on the high-affinity nerve growth factor receptor: Neuron 6:153–163

Morrell F, deToledo-Morrell L (1986) Kindling. Vol 3, New York Raven Press, pp 17–35

Phillips HS, Hains JM, Laramee GR, Rosenthal A, Winslow JW (1990) Widespread expression of BDNF but not NT-3 by target areas of basal forebrain cholinergic neurons. Science 250:290–294

Messenheimer JA, Harris EW, Steward O (1979) Sprouting fibers gain access to circuitry transsynaptically altered by kindling. Exp Neurol 64:469–481

Richardson PM, Vergelssa VMK, Riopelle RJ (1986) Distribution of neuronal receptors for nerve growth factor in the rat. J Neurosci 6:2312–2321

Rosenberg MB, Friedmann T, Robertson RC, Tuszynski M, Wolff JA, Breakefield XO, Gage F (1988) Grafting genetically modified cells to the damaged brain: restorative effects of NGF expression. Science 242:1575–1578

Rosenthal AVGD, Nguyen T, Lewis M, Shih A, Laramee GR, Nikolics K, Winslow JW (1990) Primary structure and biological activity of a novel human neurotrophic factor. Neuron 4:767–773

Schechter AL, Bothwell MA (1981) Nerve growth factor receptors on PC12 cells: evidence for two receptor classes with differing cytoskeletal association. Cell 24:867–874

Shelton DL, Reichardt LF (1984) Expression of the B-nerve growth factor gene correlates with the density of sympathetic innervation in effector organs.Proc Natl Acad Sci USA 81:7951–7955

Shelton DL, Reichardt LF (1986) Studies on the expression of the beta nerve growth factor (NGF) gene in the central nervous system: level and regional distribution of NGF mRNA suggest that NGF functions as a trophic factor for several distinct populations of neurons. Proc Natl Acad Sci USA 83:2714–2718

Scott J, Selby M, Urdea M, Quiroga M, Bell GI, Rutter WJ (1983) Isolation and nucleotide sequence of a cDNA encoding the precursor of mouse nerve growth factor. Nature 302:538–540

Siesjö BK, Bengtsson F (1989) Calcium fluxes, calcium antagonists, and calcium-related pathology in brain ischemia, hypoglycemia, and spreading depression: a unifying hypothesis. J Cereb Flow Metab 9:127–140

Soppet D, Escandon E, Maragos J, Middlemas DS, Reid SW, Blair J,Burton LE, Stanton BR, Kaplan DR, Hunter T, Nikolics K, Parada LF (1991) The neurotrophic factors brain-derived neurotrophic factor and neurotrophin-3 are ligands for the trkB tyrosine kinase receptor. Cell 65:895–903

Squinto SP, Stitt TN, Aldrich TH, Davis S, Bianco SM, Radziejweski C, Glass DJ, Masiakowski P, Furth ME, Valenzuela DM, DiStefano PS, Yancopoulos GD (1991) trkB encodes a functional receptor for brain-derived neurotrophic factor and neurotrophin-3 but not for nerve growth factor. Cell 65:885–893

Strömberg I, Wetmore CJ, Ebendal T, Ernfors P, Persson H, Olson L (1990) Rescue of basal forebrain cholinergic neurons after implantation of genetically modified cells producing recombinant NGF. J Neurosci Res 25:405–411

Sutter A, Riopelle RJ, Harris-Warrick RM, Shooter EM (1979) Nerve growth factor receptors: characterization of two distinct classes of binding sites on chick embryo sensory ganglia cells. J Biol Chem 254:5972–82

Sutula T, He XX, Cavazos J, Scott G (1988) Synaptic reorganization in the hippocampus induced by abnormal functional activity. Science 239:1147–1150

Swanson LW, Köhler C, Björklund A (1987) The limbic region. : The septohippocampal system. In: Björklund A, Hökfeld T, Swanson LW (eds). In: Handbook of chemical neuroanatomy. Vol 5: Integrad systems of the CNS, Part I. Amsterdam, Elsevier, pp 125–277

Taniuchi M, Schweitzer JB, Johson EM Jr (1986) Nerve growth factor receptor molecules in rat brain. Proc Natl Acad Sci USA 83:1950–1954

Ullrich A, Gray A, Berman C, Dull TJ (1983) Human B-nerve growth factor gene highly homologous to that of mouse. Nature 303:821–825

Westin G, Gerster T, Muller MM, Schaffner W (1987) OVEC, a versatile system to study transcription in mammalian cells and cell-free extracts. Nucleic Acids Res 15:6787–6798

Wetmore C, Ernfors P, Persson H, Olson L (1990) Localization of brain-derived neurotrophic factor mRNA to neurons in the brain by in situ hybridization. Exp Neurol 109:141–152

Whittemore SR, Ebendal T, Lörkfors L, Olsson L, Seiger A, Strömberg I, Persson H (1986) Developmental and regional expression of β-nerve growth factor messenger RNA and protein in the rat central nervous system. Proc Natl Acad Sci USA 83:817–821

Whittemore SLR, Friedman PL, Larhammer D, Persson H, Gonzalez CM, Holets VR (1988) Rat β-nerve growth factor sequence and site of synthesis in the adult hippocampus. J Neurosci Res 20:403–410

Zafra F, Hengerer B, Leibrock J, Thoenen H, Lindholm D (1990) Activity dependent regulation of BDNF and NGF mRNAs in the rat hippocampus in mediated by non-NMDA glutamate receptors. EMBO J 9:3545–3550

Regulation of the Human Neurofilament Light and Heavy Genes in Transgenic Mice

G. Charron, L. Beaudet, F. Côté, D. Houle, and *J.-P. Julien*

Summary

Using the transgenic mouse approach, we investigated the mechanisms that regulate expression of the human neurofilament light (NF-L) and heavy (NF-H) genes. Both human NF-L and NF-H genes are expressed in a tissue-specific fashion in transgenic mice. For the NF-L gene, various constructs have been tested in transgenic mice to define regulatory regions contributing to neuronal expression. Unexpectedly, this analysis demonstrated the presence of intragenic elements in the NF-L gene that confer cell-type specific regulation on the basal activity of the proximal NF-L promoter. Moreover, the examination of transgenic mice bearing a high-copy number of the NF-L gene revealed the existence of a translational control impeding overexpression of NF-L in CNS neurons. Unlike the NF-L gene, overexpression of the human NF-H gene in transgenic mice was not impeded by the existence of such posttranscriptional controls.

Introduction

A number of observations support the hypothesis that neurofilaments (NFs), the principal cytoskeletal structures of myelinated axons, play a major role during the radial growth of axons (Hoffmann et al. 1984, 1987). NFs are made by the copolymerization of three intermediate filament (IF) proteins, NF-L (61 Kd), NF-M (90 Kd) and NF-H 110 Kd), that are expressed exclusively and in the vast majority of neurons (Hoffman and Lasek 1975). The NF-L protein, which is able to self-assemble in vitro into 10-nm filaments (Geisler and Weber 1981), forms the core of the filament (Julien and Mushynski 1983). On the other hand, the NF-M and NF-H proteins that assemble in vitro into short filamentous structures (Gardner et al. 1984; Hisanaga and Hirokawa 1988) have highly charged carboxy terminals that form side-arm projections at the periphery of the NF structure (Julien and Mushynski 1983; Hisanoga and Hirokawa 1988; Geisler et al. 1983; Julien et al. 1988; Lees et al. 1988). It has been proposed that the presence of charged amino acids in the NF side-arms may induce charged repulsion that increases the spacing between the filaments and thereby contributes to the putative function of NFs in the control of axon calibers.

An understanding of the mechanisms modulating NF expression at both transcription und translation levels could shed important light on how changes in NF content occur in some pathological conditions. There is a decrease in NF levels after nerve damage (Hoffman et al. 1987), whereas an abnormal NF accumulation occurs in a variety of motor neuron diseases (Schmidt et al. 1987), in Lhermitte-Duclos disease (Yachnis et al. 1988), and in chemically induced intoxication (Parhad et al. 1988; Muma et al. 1988). Two main questions concerning NF expression have remained unanswered. First, what are the mechanisms responsible for the neuron-specific expression of NF genes? Second, how do neurons modulate the levels of NF expression?

The genes coding for the three NF proteins from various species have been sequenced and their genomic structures elucidated (Levy et al.1987; Julien et al. 1987a, 1988; Myers et al. 1987; Lees et al. 1988). Somehow, there have been inherent limitations to investigating the molecular signals that regulate the cell-type specific transcription of NF genes using in vitro approaches. For instance, no cell-type specific expression occurred when genomic clones for NF-L were transfected into cultured cells. In transfected fibroblast cell lines, the human or mouse NF-L genes were transcribed quite efficiently even though the endogenous NF-L gene remained silent (Julien et al. 1987; Monteiro and Cleveland 1989). In a study on the 5′ flanking region of the mouse NF-L gene using PC12 and C6 glioma cells, the basal promoter activity was found to be comprised within 0.3 kb upstream of the transcription start site but again no cell-specific expression was observed (Nakahira et al. 1990). A surprising result was that the NF-L promoter exhibited a very strong transcriptional activity using in vitro transcription, assay with extracts from NF-nonproducing Hela cells. These results led to the hypothesis (now refuted by our recent transgenic results) that neuronal expression in vivo of the NF-L gene may be regulated by the repression of its potent promoter in non-neuronal tissue (Nakahira et al. 1990). In any case, no DNA regions responsible for neuron-specific expression of the NF-L gene have been elucidated using in vitro systems. In contrast to results obtained with in vitro systems, DNA fragments bearing either the complete human or mouse NF-L genes turned out to be correctly regulated in transgenic mice (Julien et al. 1987b; Monteiro et al. 1990). The suitability of the transgenic mouse approach to address the problem of NF gene regulation prompted us to investigate the mechanisms controlling the cell-type specificity as well as the levels of NF-L and NF-H expression.

Results and discussion

Intragenic elements contribute to NF-L transcription

An analysis of the mouse NF-L promoter by transient expression assays with cultured cells indicated that the basic transcription activity of the NF-L 5′ region was comprised within 0.3 kb (Nakahira 1990). To assess in vivo the activity of the human NF-L promoter, a 0.3 kb fragment was fused to the chloramphenicol acetyl transferase (CAT) and introduced into the mouse germ line. We have analyzed the levels of CAT activity in various tissues of mice from two transgenic lines (data not shown). The CAT activity was detected at low levels in all tissues examined. We concluded that the proximal NF-L 5′ region constitutes a weak promoter and that it lacks regulatory elements to direct neuron-specific expression. Clearly, this 5′ flanking NF-L region is not sufficient to account for the full activity of the NF-L gene. The presence of TATAA, Sp1 and AP-1-like elements within this region may be responsible for this basal transcription activity (Fig. 1).

Figure 2 summarizes the results with various NF-L DNA fragments that were tested in transgenic mice. Surprisingly, correct mRNA expression occurred

```
BamH1
GGATCCAAGT GTCACGGGGT CTGGGCAATG CAGGACGGGA GGGGCTGCGT GAGTGAGTAC AGAAGGGAAA

TGAGTGAGGG GGCATGGGAT CTCAGAGAAA ATCAGGGACC TCTGAGCAAA GTGGAAAGGA CGACCGCCGC

                      Sp1
AGCTCCTCGG GCCGTAGCTC GACCCCGCCT TCCCTTTTGC GCAGAATCCT CGCCTTGGCT GCAGCAGCGC

                                        AP1                           Tata box
GCTGCCCCCA CTGGCCGGCG TGCCGTGATC GATCGCAGGC TGCGTCAGGA GCCTCCCGGC GTATAAATAG

                     Capping site
GGGTGGCAGA ACGGCGCCGA GCCGCACACA GCCATCCATC CTCCCCCTTC CCTCTCTCCC CTGTCCTCTC

TCTCCGGGCT CCCACGCCGC CGCGGCCGGG GAGCCACCGG CCGCCACCAT GAGTTCCTTC AGCTACGAGC

                                        Oligo
CGTACTACTC GACCTCCTAC AAGCGGCGCT ACGTGGAGAC GCCCCGGGTG CACATCTCCA GCGTGCGCAG

CGGCTACAGC ACCGCACGCT CA
```

Fig. 1. Sequence of the promoter region of the human NF-L gene. The putative regulatory elements are marked. An oligonucleotide complementary to the underlined coding sequence has been used for hybridization of the Northern blots shown in Figure 2.

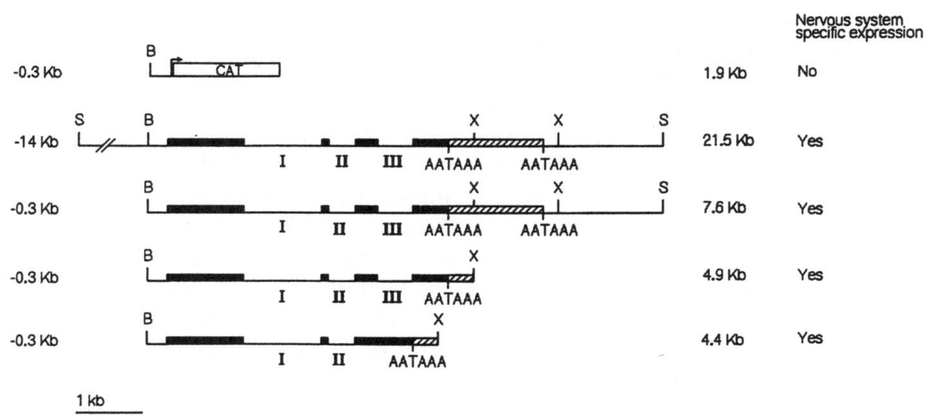

Fig. 2. Activity of various NF-L constructs in transgenic mice.

when the basal NF-L promoter (–0.3 kb) was combined with NF-L sequences dowstream of the mRNA start site. The 5′- and 3′-NF-L sequences are largely dispensable for neuronal expression. Thus, a BamHI/XbaI NF-L fragment of 4.9 kb in size that included only the first polyadenylation site was sufficient for tissue-specific expression. Our conclusion from this analysis is that intragenic regulatory elements must contribute to transcriptional control of the NF-L gene. The existence of internal regulatory elements was unexpected. For most genes, including members of the IF gene family, the cell-type specificities appear to be controlled by elements located in the 5′ flanking region of the genes (Pieper et al. 1987; Krimperfort et al. 1988; Blessing et al. 1989, Leask et al. 1990). An enhancer has been found in the first intron of the keratin 18 gene but it acts as a non-cell-type specific enhancer (Oshima et al. 1990). A NF-L DNA transgene of 4.4 kb in which the third intron has been removed was still expressed in a tissue-specific fashion. Thus, intron 1 and/or intron 2, which are unique to members of the NF gene family, may contain NF-L regulatory elements required for neuronal expression. Work is now in progress with additional NF-L constructs to further define the contribution of these introns in modulating neuron-specific expression. It is our hope that characterization of NF-L regulatory elements might help to design promoter/enhancer elements capable of targeting high level foreign gene expression to neurons.

Different post-transcriptional controls for the NF-L and NF-H genes

In an attempt to overexpress NF-L proteins, we generated a transgenic mouse line that carries approximately 500 copies of a 21.5 kb human NF-L fragment (Julien et al. 1987b). Transgenics from this line (no. 29) produce up to three times the levels of NF-L mRNAs in the central nervous sytem (CNS), as illustrated on the Northern blot in Figure 3A. Both human and mouse NF-L mRNAs were detected by hybridizing 10 µg of RNA samples from spinal cord with a 30 mer oligonucleotide that spans a conserved coding sequence of the N-terminal regions. The considerable increase in NF-L mRNAs did not trigger a corresponding increase in the levels of NF-L proteins, as revealed by an analysis on SDS gels of spinal cord samples from normal and transgenic mice (Fig. 3B). Thus, densitometry scanning of Coomassie blue stained gels showed only 50% increase in the levels of NF-L proteins in transgenic spinal cord samples having three times the levels of NF-L mRNA. The human NF-L proteins comigrate with the endogenous mouse NF-L proteins on SDS gels. Nonetheless, according to experiments based on intensities of Coomassie blue staining and immunodetection with a specific anti-human NF-L antibody (monoclonal DP5-112 derived by D. Paulin, Paris), we estimate that human NF-L species constitute approximately 80% of total NF-L content (data not shown). We conclude that the post-transcriptional regulation must affect human and mouse NF-L mRNA species evenly. Evidence for post-transcriptional mechanisms regulating mouse NF-L expression in the CNS has been previously provided by the analysis of a MSV promoter/mouse NF-L fusion construct in transgenic mice (Monteiro et al. 1990), although in this latter situation the phenomenon

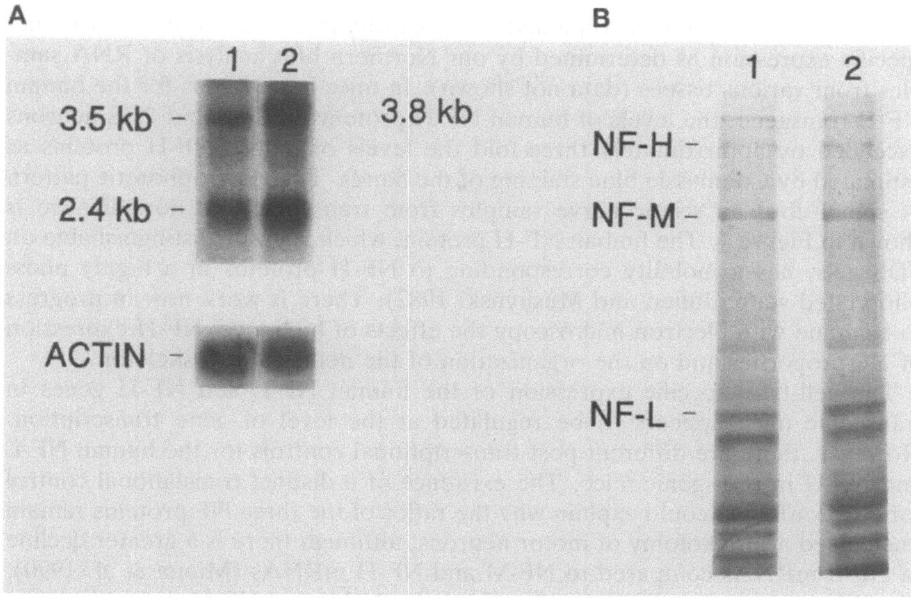

Fig. 3A–B. Analysis of human NF-L mRNA and protein in transgenic mice with a high-copy number of the human NF-L gene. *A*, the human and mouse NF-L mRNA species were detected by Northern blotting using a 30 mer oligonucleotide probe spanning a conserved sequence of the coding region (see Fig. 1). Ten micrograms of total RNA from the spinal cords of normal (1) or transgenic (2) mice were used for hybridization. The human NF-L transcripts are 2.4 and 3.8 kb in size; the corresponding mouse transcripts are 2.4 and 3.5 kb. A 3-fold increase in NF-L mRNA species occurs in the spinal cords of transgenic mice. *B*, SDS gels of total protein from normal (1) and transgenic (2) spinal cords. The spinal cords were homogenized in sample buffer and electrophoresed on a 7.5% polyacrylamide SDS gel. The bands are stained with Coomassie blue. A 50% increase in NF-L proteins resulted from a 3-fold increase in mRNAs.

might have been the result of changes introduced in the 5′ untranslated region of the mouse NF-L fusion gene (Monteiro et al. 1989). To further investigate whether overexpression of NF-L proteins in CNS neurons was being defeated by translational or post-translational events, we carried out in vivo labeling experiments in the spinal cords of anesthetized mice with ^{35}S-methionine to determine the levels of newly synthesized NF-L proteins. After a period of 5, 15 or 60 min, there was no significant increase in the amount of ^{35}S-labeled NF-L proteins in spinal cord samples of transgenic mice when compared to samples from normal mice (data not shown). Thus, the results provided evidence for the existence of translational controls for NF-L expression.

In contrast to the situation with NF-L, overexpression of human NF-H was not impeded by the existence of post-transcriptional controls. Recently, we generated transgenic mouse lines bearing multiple copies of a 40 kb genomic fragment for the human NF-H gene. A cosmid clone for human NF-H was isolated from a library enriched for chrosomome 22. The genomic clone was linearized with *Not*I restriction enzyme and introduced into the mouse germ line. This NF-H fragment contained all regulatory elements required for tissue-specific expression as determined by our Northern blot analysis of RNA samples from various tissues (data not shown). In mice hemizygous for the human NF-H transgene, the levels of human NF-H proteins in CNS and PNS neurons exceeded by approximately three-fold the levels of mouse NF-H proteins as estimated by Coomassie blue staining of the bands. The electrophoretic pattern of spinal cord an sciatic nerve samples from transgenic and normal mice is shown in Figure 4. The human NF-H protein, which is easily distinguishable on SDS gels, has a mobility corresponding to NF-H proteins in a highly phosphorylated state (Julien and Mushynski 1982). There is work now in progress to examine with electron microscopy the effects of high-level NF-H expression of NF properties and on the organization of the neuronal cytoskeleton.

The cell-type specific expression of the human NF-L and NF-H genes in transgenic mice appears to be regulated at the level of gene transcription. However, there are different post-transcriptional controls for the human NF-L and NF-H in transgenic mice. The existence of a distinct translational control for NF-L mRNAs could explain why the ratios of the three NF proteins remain unchanged after axotomy of motor neurons, although there is a greater decline of NF-L mRNAs compared to NF-M and NF-H mRNAs (Muma et al. 1990). The different post-transcriptional controls for NF-L and NF-H expression may reflect the different roles of the two proteins in the neuronal function. As mentioned above, the NF-L protein forms the core of the filament (Geisler and Weber 1981; Julien and Mushynski 1983) whereas NF-H proteins form sidearm projections at the periphery of the NF structure (Julien and Mushynski 1983; Geisler et al. 1983; Hisanaga and Hirokawa 1988). Perhaps it is critical for neurons to finely regulate the levels of NF-L proteins that have the competence to polymerize into 10 nm filaments. An autoregulation mechanisms has been described previously for another cytoskeletal protein, the beta-tubulin

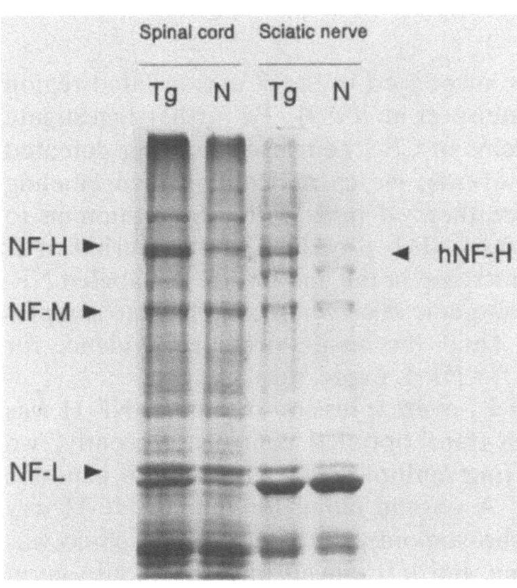

Fig. 4. Overexpression of human NF-H proteins in transgenic mice. Total protein from the spinal cord an sciatic nerve of transgenic and normal mice was analyzed on SDS gels as described in Figure 3. The human NF-H proteins have a slightly higher mobility on SDS gels.

(Pachter et al. 1987). This autoregulation occurs through sequences within exon 1 of beta-tubulin mRNA that modulate mRNA stability. However, for the NF-L post-transcriptional control described above, the mechanism is not established through changes in mRNA stability, as discrete bands are detected on Northern blots (Fig. 3A). Another mechanism could involve a negative translational control of mRNAs by the specific binding of a regulatory protein to a mRNA sequence impeding the migration of the ribosomes down the message, such as the control of ferritin by an iron-responsive element (Casey et al. 1988; Hershey 1991). Future experiments aiming to test various NF-L mutant constructs in transgenic mice should help in the elucidation of the NF-L mRNA region involved in this intriguing translation control.

Acknowledgments. This work has been supported by the Medical Research Council of Canada and the Canadian Centre of Excellence on Neural Regeneration.

References

Blessing M, Jorcano JL, Franke WW (1989) Enhancer elements directing cell-type-specific expression of cytokeratin genes and changes of the epthelial cytoskeleton by transfections of hybrid cytokeratin genes. EMBO J 8:117–126

Casey JL, Hentze MW, Koeller DM, Caughman SW, Rouault TA, Klausner RD, Harford JB (1988) Iron-responsive elements: Regulatory mRNA sequences that control mRNA levels and translation. Science 240:924–928

Clark AW, Krekowski CA, Parhad IM, Liston D, Julien J-P, Hoar DI (1989) Altered expression of genes for amyloid and cytoskeletal proteins in Alzheimer cortex. Ann Neurol 25:331–339

Gardner EE,Dahl D, Bignami A (1984) Formation of 10 nm filaments from the 150 kd neurofilament protein in vitro. J Neurosci Res 11:145–155

Geisler N, Weber K (1981) Self-assembly in vitro of the 68,000 molecular weight component of the mammalian neurofilament triplet proteins into intermediate filament-sized filaments. J Mol Biol 151:565–571

Geisler N, Kaufmann E, Fisher S, Plessmann U, Weber K (1983) Neurofilament architecture combines structural principles of intermediate filaments with carboxy-terminal extensions of increasing size between triplet proteins. EMBO J 2:1295–1302

Graham DG, Szakal-Quin G, Priest JW, Anthony DC (1984) In vitro evidence that covalent crosslinking of neurofilaments occurs in gamma-diketone neuropathy. Proc Natl Acad Sci USA 81:4979–4982

Hershey JWB (1991) Translational control in mammalian cells. Annu Rev Biochem 60:717–755

Hisanaga S, Hirokawa N (1988) Structure of the peripheral domains of neurofilaments revealed by low angle rotatory shadowing. J Mol Biol 202:297–305

Hoffman PN, Cleveland DW (1988) Neurofilament and tubulin expression recapitulates the developmental program during axonal regeneration. Proc Natl Acad Sci USA 85:4530–4533

Hoffman PN, Cleveland DW, Griffin JW, Landes PW, Cowan NJ, Price DL (1987) Neurofilament gene expression: A major determinant of axonal caliber. Proc Natl Acad Sci USA 84:3472–3476

Hoffman PN, Griffin JW, Price DL (1984) Control of axon caliber by neurofilament transport. J Cell Biol 99:705–714

Hoffman PN, Lasek RJ (1975) The slow component of axonal transport. Identification of major structural polypeptides of the axon and their generality among mammalian neurons. J Cell Biol 66:351–366

Julien J-P, Mushynski WE (1983) The distribution of phosphorylation sites among identified proteolytic fragments of mammalian neurofilaments. J Biol Chem 258:4019–4025

Julien J-P, Grosveld F, Yazdanbaksh K, Flavell D, Meijer D, Mushynski W (1987a) The structure of a human neurofilament gene (NF-L): a unique exon-intron organization in the intermediate filament gene family. Biochim Biophys Acta 909:10–20

Julien J-P, Tretjakoff I, Beaudet L, Peterson A (1987b) Expression and assembly of a human neurofilament protein in transgenic mice provide a novel neuronal marking system. Genes Dev 1987:1085–1095

Julien J-P, Côté F, Beaudet L, Sidky M, Flavell D, Grosveld F, Mushynski WE (1988) Sequence and structure of the mouse gene coding for the largest neurofilament subunit. Gene 68:307–314

Krimperfort PJ, Schaart G, Pieper FR, Ramaekers FC, Guypers HT, Van den Neuvel RM, Vree Egberts WT, Van Eys GY, Berns A, Bloemendal H (1988) Tissue-specific expression of a vimentin-desmin hybrid gene in transgenic mice. EMBO J 7:941–947

Leask A, Rosenberg M, Vassar R, Fuchs E (1990) Regulation of a human epidermal keratin gene: sequences and nuclear factors involved in keratinocyte-specific transcription. Genes Dev 4:1985–1998

Lees JF, Shneidman PS, Skuntz SF, Carden MJ, Lazzarini RA (1988) The structure and organization of the human heavy neurofilament subunit (NF-H) and the gene encoding it. EMBO J 7:1947–1955

Levy E, Liem RKH, D'Eustachio P, Cowan NJ (1987) Structure and evolutionary origin of the gene encoding mouse NF-M, the middle-molecular-mass neurofilament protein. Eur J Biochem 166:71–77

Monteiro MJ, Cleveland DW (1989) Expression of NF-L and NF-M in fibroblasts reveals coassembly of neurofilament and vimentin subunits. J Cell Biol 108:579–593

Monteiro MF, Hoffman PN, Gearhart JD, Cleveland DW (1990) Expression of NF-L in both neuronal and nonneuronal cells of transgenic mice: increased neurofilament density in axons without affecting caliber. J Cell Biol 111:1543–1557

Muma NA, Hoffman PN, Slunt HH, Applegate MD, Lieberburg I, Price DL (1990) Alterations in levels of mRNAs coding for neurofilament protein subunits during regeneration. Exp Neurol 107:230–235

Muma NA, Troncoso JC, Hoffman PN, Roo EH, Price DL (1988) Aluminium neurotoxicity: Altered expression of cytoskeletal genes, Mol Brain Res 3:115–122

Myers MW, Lazzarini RA, Lee VM-Y, Schlaepfer WW, Nelson DL (1987) The human mid-size neurofilament subunit: a repeated protein sequence and the relationship of its gene to the intermediate filament gene family. EMBO J 6:1617–1626

Nakahira K, Ikenaka K, Wada K, Tamura T, Furuichi T, Mikoshiba K (1990) Structure of the 68-kda neurofilament gene and regulation of its expression. J Biol Chem 265:19786–19791

Oshima RG, Abrams L, Kulesh D (1990) Activation of an intron enhancer within the keratin 18 gene by expression of c-fos and c-jun in undifferentiated F9 embryonal carcinoma cells. Genes Dev 4:835–848

Pachter JS, Yen TJ, Cleveland DW (1987) Autoregulation of tubulin expression is achieved through specific degradation of polysomal tubulin mRNAs. Cell 51:283–292

Parhad IM, Swedberg EA, Hoar DI, Kerkoski CA, Clark AW (1988) Neurofilament gene expression following iminodipropionitrile (IDPN) intoxication. Mol Brain Res 4:293–301

Pieper FR, Slobbe RL, Ramaekers FCS, Cuypers HT, Bloemendal (1987) Upstream regions of the hamster desmin and vimentin genes regulate expressing during in vitro myogenesis. EMBO J 6:3611–3618

Schmidt ML, Carden MJ, Lee VM-Y, Trojanowski JQ (1987) Phosphate dependent and independent neurofilament epitopes in the axonal swellings of patients with motor neuron disease and controls. Lab Invest 56:282–294

Yachnis AT, Trojanowski JQ, Memmo M, Schlaepfer WW (1988) Expression of neurofilament proteins in the hypertrophic granule cells of Lhermitte-Duclos disease: An explanation for the mass effect and the myelination of parallel fibers in the disease state. J Neuropathol Exp Neurol 47:206–216

Subject Index

MIX
Papier aus verantwortungsvollen Quellen
Paper from responsible sources
FSC® C105338

If you have any concerns about our products,
you can contact us on
ProductSafety@springernature.com

In case Publisher is established outside the EU,
the EU authorized representative is:
Springer Nature Customer Service Center GmbH
Europaplatz 3, 69115 Heidelberg, Germany

Printed by Libri Plureos GmbH
in Hamburg, Germany